Drug Delivery Applications of Starch Biopolymer Derivatives

Jin Chen · Ling Chen · Fengwei Xie ·
Xiaoxi Li

Drug Delivery Applications of Starch Biopolymer Derivatives

Springer

Jin Chen
Guangdong Province Key Laboratory for Green Processing of Natural Products and Product Safety, School of Food Science and Engineering, Ministry of Education Engineering Research Center of Starch and Protein Processing
South China University of Technology
Guangzhou, Guangdong, China

Ling Chen
Guangdong Province Key Laboratory for Green Processing of Natural Products and Product Safety, School of Food Science and Engineering, Ministry of Education Engineering Research Center of Starch and Protein Processing
South China University of Technology
Guangzhou, Guangdong, China

Fengwei Xie
International Institute for Nanocomposites Manufacturing (IINM), WMG
University of Warwick
Coventry, UK

Xiaoxi Li
Guangdong Province Key Laboratory for Green Processing of Natural Products and Product Safety, School of Food Science and Engineering, Ministry of Education Engineering Research Center of Starch and Protein Processing
South China University of Technology
Guangzhou, Guangdong, China

ISBN 978-981-13-3656-0 ISBN 978-981-13-3657-7 (eBook)
https://doi.org/10.1007/978-981-13-3657-7

Library of Congress Control Number: 2018964237

This Springer imprint is published by the registered company Springer Nature Singapore Pte Ltd.
The registered company address is: 152 Beach Road, #21-01/04 Gateway East, Singapore 189721, Singapore

Acknowledgements

This book has been financially supported by the National Key R&D Program of China (No. 2016YFD0400203), the NSFC (31871751), the Key Project of Guangzhou Science and Technology Program (No. 201804020036) and YangFan Innovative and Entrepreneurial Research Team Project (2014YT02S029).

Contents

Nomenclature

10-HCPT	10-hydroxy camptothecin
AA	Acrylic acid
Apo	Apolipoprotein
BMP-4	Bone morphogenetic protein-4
BSA	Bovine serum albumin
CMC	Critical micelle concentration
CMS	Carboxymethyl starch
Con A	Concanavalin A
DC	Dendritic cell
DDS	Drug delivery system
DMAAm	*N,N*-dimethylacrylamide
DOX	Doxorubicin
DS	Degree of substitution
DSM	Degradable starch microspheres
EMF	External magnetic field
EPR	Enhanced permeability and retention
FA	Folic acid
GIT	Gastrointestinal tract
GSH	Glutathione
HA	Hyaluronic acid
HAase	Hyaluronidase
HACS	High-amylose corn starch
HAS	Human serum albumin
HEMA	Hydroxyethyl methacrylate
HES	Hydroxyethyl starch
HPS	Hydroxypropyl starch
IC_{50}	Half maximal inhibitory concentration
IL	Interleukin
IPN	Interpenetrating polymer network
KC	Kupffer cell

LCST	Low critical solution temperature
LPC	Lysophosphatidylcholine
M	Membranous
M	Molecular weight
MMT	Montmorillonite
MNP	Magnetic nanoparticle
MP	Microparticle
mPEG	Methoxypolyoxy-ethylene amine
MS	Molar degree of substitution
N/P	Nitrogen/phosphate atom
NIR	Near-infrared region
NP	Nanoparticle
PAA	Polyacrylic acid
PAN	Polyacrylonitrile
PDMAEM	Poly(*N*,*N*-dimethylaminoethyl methacrylate)
PEC	Polyelectrolyte complex
PEG	Poly(ethylene glycol)
PEI	Poly(ethyleneimine)
PMAA	Polymethacrylic acid
PMMA	Poly(methyl methacrylate)
PNIPAM	Poly(*N*-isopropylacrylamide)
PS80	Polysorbate 80
RES	Reticuloendothelial system
ROS	Reactive oxygen species
RS	Resistant starch
SA	Starch acetate
SCF	Simulated colon fluid
SGF	Simulated gastric fluid
SIF	Simulated intestinal fluid
TNF	Tumor necrosis factor
UCST	Upper critical solution temperature
VPTT	Volume phase transition temperature
ZO	Zonula occluden

Overview

Being cheap, renewable, biodegradable, and biocompatible, starch has attracted huge interests from drug delivery scientists. Recently, the application of starch in drug delivery systems (DDSs) has made significant advances, whereby the molecular structure and characteristics of starch are exploited for creating smart materials for drug delivery purposes. With a better understanding of the physiological conditions, the difference between normal and pathological cells, and the various biological interactions between materials and physiological structures, starch-based DDSs with stimulus responsiveness (e.g., pH, temperature, or redox potential), target specificity, and bioadhesiveness can be constructed. These new DDSs would be capable of addressing some of the systemic and intracellular barriers in drug delivery and thus enhancing the bioavailability of therapeutics at the disease sites. This monograph highlights recent emerging areas in the design of starch-based materials with advanced drug delivery behaviors. Key design principles, challenges, and prospects of these starch-based DDSs are also discussed.

Chapter 1
Physiological and Pathological Bases for Designing High Performance Drug Delivery Carriers

Abstract The strategy of using drug delivery systems (DDSs) to deliver drugs will confront different physiological environments in the human body. A better understanding of the physiological base is the key step for the success of advanced starch-based DDSs with high performance. For example, various biological stimuli, such as temperature, pH, and hypoxia at the organ, tissue, and cell levels can be exploited to trigger the stimulus-responsive delivery of therapeutics. Besides, external stimuli such as electric/magnetic fields or light have been developed to construct stimulus responsiveness of drug carriers. On the other hand, when it comes to clinical application, DDS will confront different physiological environments along with various administration routes, such as oral cavity mucosa or blood circulation. The drug carriers will interact with these physiological environments at tissues, cell, and molecular levels. With a deeper understanding of these physiological bases, further advances have been introduced by DDSs endowed with target-specific and bioadhesive properties to various sites in the body to reduce the side effects and enhance the bioavailability of drugs at the diseased site. Regarding this, the physiological and pathological hallmarks of the organ or cell levels and the interactions between the physiological structure and materials are highlighted in this section.

Keywords Physiological and pathological bases · Biological stimuli · Physical stimuli · Interactions with mucosa · Interactions with cells · Interactions with biomacromolecules

1.1 Concepts for Designing Stimulus-Responsive Carriers

The examples of the stimuli that can be exploited for triggering the delivery of drugs, or diagnostic agents from the carriers are diverse and manifold in nature (Fig. 1.1). The variation in physiological and pathological parameters of organ or cell levels is often important hallmarks for designing stimulus-responsive materials, especially for some distinct types of diseases with abnormal signs, such as cancer, infections and cardiovascular diseases (Hoffman 2013). These stimuli are naturally occurring in vivo and can be classified as internal stimuli (physiological, pathological, and

J. Chen et al., *Drug Delivery Applications of Starch Biopolymer Derivatives*
https://doi.org/10.1007/978-981-13-3657-7_1

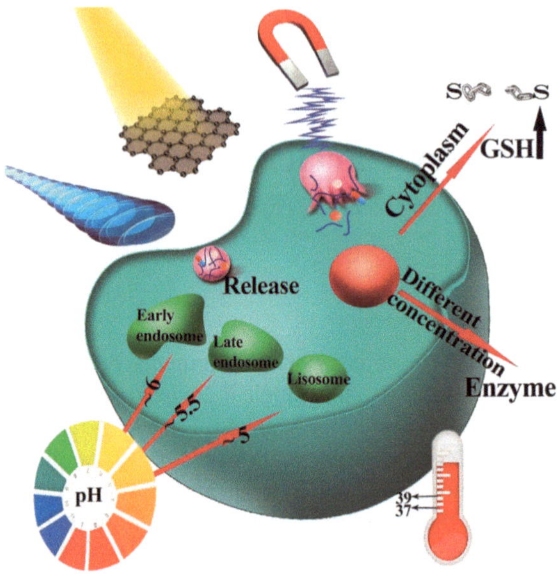

Fig. 1.1 Examples of different internal and physical stimuli exploited for triggering carriers in controlled release of drugs, namely, magnetic induction, light irradiation, ultrasound agitation, temperature difference, enzyme activation, redox potential [oxidation–reduction of glutathione (GSH)], or pH difference. Reproduced from reference Karimi et al. (2016), Copyright (2016) with permission from the Royal Society of Chemistry

pathochemical conditions). In contrast, those stimuli (e.g., temperature, light, and magnetic and electrical fields) that can be applied externally and artificially can be classified as physical stimuli.

1.1.1 Biological Stimuli

Internal stimuli of biochemical and biologic origin include pH shift, enzyme, redox microenvironment, and some biomolecules of the specific tissues, or in certain pathological states (Fig. 1.1). In general, these stimuli can be classified into three main categories: triggers at the organ level, triggers related to pathological conditions, and cellular compartment-specific triggers. In some instances, some stimuli are overlapped, such as pH shift and enzyme.

The pH shift is found in specific tissues and cellular component, and the normal pH gradient can be affected by a diseased state. The obvious change in pH along the human gastrointestinal tract (GIT) has to be considered for oral delivery of any kind of drug. Table 1.1 summarizes some of the critical processes and parameters in different sections of the human GIT. Apart from the stomach, the remainder of the GIT is generally in the pH range of 5–8. The pH tends to rise gradually from the duodenum and reaches a peak just before entry into the colon. After that the pH falls in the proximal colon and gradually rises again in the distal colon (Watts and Smith 2005). While at the cellular level, the pathological state can affect the normal pH gradient existing between the extra and intracellular environments (Karimi

Table 1.1 Segments of the gastrointestinal tract and their corresponding functions

Organ	Main processes	Secretions/enzymes	pH
Mouth	Mechanical digestion of food (chewing, grinding)	Amylase (in saliva)	–
Esophagus	Passage of food from mouth through stomach	–	–
Stomach	Storage and mechanical digestion of food start of chemical digestion (pepsinogen secretion and activation); Reduction in the number of bacteria	Pepsinogen, which transformed into pepsin under acid conditions	1–3 fasted 4–7 fed
Small intestine	Continuation and completion of chemical digestion of carbohydrates, protein, and lipids Absorption of small soluble nutrients	Pancreatic juice Intestinal juice Bile	pH gradient along the small intestine ~5–7.5
Large intestine (colon)	Water absorption from indigestible feces (proximal) (distal)	–	Proximal 5–7 Distal ~7

Reprinted from reference Sandra (2009), Copyright (2009), with permission from the American Chemical Society

et al. 2016). In solid tumors, the extracellular pH can be significantly more acidic (~6–7) than systemic pH (7.4) due to the poor vasculature and consequent anaerobic acid conditions prevailing in the malignant cells. Besides, the cellular organelles also exhibit sharp pH differences in different locations. Particles entering cells on endocytic pathways very likely experience a drop in pH from neutral to pH 5.9–6.0 in the lumen of early/recycling vesicles, with a further reduction during progression from late endosomes (5.0–6.0) to lysosomes (4.5–5.0) (Grabe and Oster 2001).

Owing to the varied roles that enzymes have in different biological processes, enzymes are widespread in specific tissues and cellular component. Enzyme responsiveness can be an excellent strategy for designing smart materials, especially for degradable polymers. When the oral administration is practiced, digestive enzymes of diverse specificities are found in saliva, stomach, pancreatic juice, and the small and large intestines along the GIT (Bernkop-Schnürch 1998). Another digestion part that has interested scientists is the gastrointestinal bacterial population (Table 1.1).

These bacteria can ferment polymers that are not digested by the enzymes present in the human GIT (Yang et al. 2002). The pH and motility of the large intestine are much more conducive to bacterial growth (10^{11}–10^{12} CFU/g contents) than the upper intestine (10^4–10^7 CFU/g) and the stomach (10^2 CFU/g) (Bernhardt and Knoke 1997; Simon and Gorbach 1984). And, the colonic microbiota is said to have a metabolic potential equal to, or greater than, that of the liver (Scheline 1973). Besides, like pH shift, particles will be encapsulated in the lysosome, which is an organelle with high activity of hydrolases, such as lysosomal hydrolases (Jovic et al. 2010). Moreover, the diseased state can also affect the overexpressed enzymes and their concentration may become higher. An array of manifold enzymes, such as proteases, glucuronidase, or carboxylesterases, are either intra- or extracellular presented by normal and malignant cells and can be used as well as a biochemical trigger (Yang et al. 2011).

Differential reductive environments present in pathological states, particularly in the tumor microenvironment, can also trigger and thus facilitate drug release with the good response to the high intracellular levels of GSH (Meng et al. 2012). Research (Wang et al. 2013) has shown that the intracellular environment of reductive GSH is ~10 mM, which is significantly higher than the concentration in the extracellular environment (~2 μM). Also, the endosomal compartment is redox-active due to the presence of lysosomal thiol reductase (Karimi et al. 2016). Furthermore, the concentration of GSH and the cytosol of tumor cells are higher than that in normal tissues. A high redox potential difference (~100–1000 folds) exists between the reducing intracellular and oxidizing extracellular spaces in some tumor cells (Saito et al. 2003). The combination of intracellular-elevated GSH and the tumor-associated GSH make redox-responsive nanocarriers interesting candidates for targeted drug release.

The existence of specific biomolecules such as glucose, reactive oxygen species (ROS), and nucleic acids in specific physiological sites or in pathological conditions has encouraged the development of biomolecule-sensitive carriers. For example, glucose responsiveness, based on the conversion of glucose to gluconic acid by glucose oxidase (Lu et al. 2014), have been most extensively investigated to control the insulin therapy for diabetics in response to blood glucose levels; ROS responsiveness can be designed to operate based on ROS degradation, such as HO^-/H_2O_2, which are produced in the cell as a by-product of electron transfer reactions (Lee et al. 2013). ROSs are associated with distinct pathological conditions including cancer, stroke, arteriosclerosis, and tissue injury (Devasagayam et al. 2004). Moreover, due to the various roles in different biological processes and the unique hybridization features, nucleic acids, including RNA and DNA, have recently emerged as important biological triggers, especially for microRNAs. MicroRNAs have been revealed to be involved in the initiation and progression of cancer (Ventura and Jacks 2009). With the ability to fold into secondary structures, aptamers, nucleic acids can specifically recognize and bind small molecule drugs. Consequently, drug-nucleic acid mimics can be used to recognize cancer cells "pre-targeted" with antibodies carrying complementary nucleic acids (Liu et al. 2003).

1.1.2 Physical Stimuli

Physical stimuli that can be applied to bring about a triggered release of bioactives may involve temperature, light, ultrasound, mechanical force, and magnetic or electrical fields (Mura et al. 2013). Physical stimuli can be obtained both internally or externally, e.g., temperature. Temperature sensitivity is one of the most interesting characteristics in stimulus-responsive polymeric carriers and has been extensively investigated to exploit the pathological temperature of diseases, especially for cancer treatment (Meyer et al. 2001; Rijcken et al. 2007). The hyperthermic nature of most inflamed pathological sites and tumors can act as an internal stimulus (Torchilin 2000). Also, applying external temperature changes to activate thermoresponsive materials provides another option. Utilization of external stimuli usually offers a range of advantages, including the ease of application, relative biocompatibility, and controllability both spatially and temporally (Han et al. 2006; Koçer et al. 2005).

The utilization of light irradiation as stimulus is tissue compatible, yet powerful enough to bring about conformational changes within the architecture of designed carriers. Ultraviolet (10–400 nm), visible, or near-infrared regions (NIRs) (650–900 nm) of the light spectrum can be used to trigger drug release from carriers (Wang et al. 2012). Yet, when using the light spectrum as stimuli, their high energy may cause damage to normal cells, while low energy hinders their penetration into the body. Regarding this, NIR has been the most often investigated stimuli to control the release of drugs from stimulus-responsive DDSs among these light irradiations. Although the low energy of NIR photons, NIR has better transmission through tissues due to its lower absorption and scattering in tissues (penetrating into the body about 10 cm), due to its bare absorption of hemoglobin, water, and lipids (Yi and Sukhorukov 2014).

During the last decade, the mechanical force has attracted growing attraction in the field of responsive DDSs, due to the facility of treatment in a variety of conditions with convenient commands (Zhang et al. 2016). A wild range of mechanical force is ubiquitously achieved in the body or easily applied externally. To be specific, the compressive and tensile forces can be readily obtained by using hands or from simple daily motions, such as tension in muscles, tendons, and bone joints, as well as compression in cartilage and bones (Guskiewicz and Mihalik 2011). Besides, along with the GIT, increased pressure exists in the stomach and large intestine, due to the increased viscosity caused by the food in the stomach (Arora et al. 2005) or the reabsorption of water from the large intestine (Ritschel 1991). While the shear force can be generated by blood flow in the human vascular system, obstruction of normal blood flow results in a dramatic increase in shear stress (1–2 orders of magnitude), which poses a threat to human health in the forms of stroke and atherosclerosis (Nesbitt et al. 2009). Specifically, shear stress in normal vessels generally remains below ~70 dyn/cm^2, while a 95% constricted stenotic or thrombosed artery possesses stress above 1000 dyn/cm^2 (Bark and Ku 2010; Wootton and Ku 1999).

Magnetic activation-based approaches are more attractive from a biotechnological point of view because the magnetic field does not affect most biological tissues, compared to other stimuli such as light irradiation. The application of magnetic nanoparticles (MNPs) and external magnetic field (EMF) can provide a real-time triggering impulse and biomedical targeting. MNPs consist of a metal or metal oxide core, encapsulated in an inorganic or a polymeric coating that renders the particles with biocompatibility and stability, and may serve as a support for biomolecules. Upon administration, the drugs immobilized in magnetic carriers can be accumulated at targeted sites under the direction of EMF through strong enough magnetic force generated on NPs and avoid the clearance by the reticuloendothelial system (RES) (also known as the mononuclear phagocytic system) (Likhitkar and Bajpai 2014). To ensure a strong magnetic force generated on NPs, MNPs with larger magnetic cores (>100 nm) are of great importance for their selective delivery to tumors via magnetic targeting (Cole et al. 2011). A large particle size, however, can attenuate the residence time of MNPs in the blood circulation and can cause severe MNPs aggregation, which results in significant embolism of the afferent vasculature and further leads to severe neurosequelae (Cole et al. 2011). Poly(ethylene glycol) (PEG) is the most commonly used polymer to stabilize NPs in the blood serum because of the hydrophilic and flexible PEG segments. Besides, magnetic fluid hyperthermia is another field where MNPs are applied to create an increase in temperature by applying an oscillating magnetic field to kill tumor cells (Laurent et al. 2011).

Similar to magnetic-field-responsive carriers, electroresponsive carriers have been investigated to control the release of drugs from stimulus-responsive DDSs with a spatiotemporal control. Much of the field has focused on electroresponsive hydrogels, and the large water content of hydrogels (up to 99%) allows for efficient transport of electrical ions in response to an electric field. The main mechanism by which drugs are released in these systems involves the forced convection of drug out of the gel as it de-swells in response to imposed electric fields (Murdan 2003).

In recent years, ultrasound as a highly efficient and noninvasive trigger has been widely explored in the drug delivery field. The longitudinal force produced by ultrasound can induce hyperthermia, cavitation, and radiation, releasing active molecules from the polymeric matrix by regional sonication with the aid of microbubbles (Gao et al. 2005).

1.2 Interactions Between Carriers and Physiological Structures

Different routes of administration of DDSs, such as oral, transmucosal, and parenteral (systemic and local) routes, will confront different physiological environments in clinical application. Determining how carriers interact with these physiological structures varying different routes of administration are essential. In addition, interactions between tissues, cells or biomacromolecules and materials are dramatically important for the better performance of DDSs. Regarding this, the interactions of carriers with mucosa through transmucosal route, and those with cells and biomacromolecules, that may influence in vivo circulation and tissue targeting, are highlighted here.

1.2.1 Interactions with Mucosa

The mucosal layer exists in some regions of the body including the GIT, the urogenital tract, the airways, the ear, nose, and eye. Mucosa is composed of one or more layers of epithelial cells and an underlying loose connective tissue. For example, the oral mucosa is composed of an outermost layer of stratified squamous epithelium, a basement membrane and a lamina proprium followed by the submucosa as the innermost layer (Harris and Robinson 1992). The type of cells varies from organ to organ and can differ along a given tract of the same organ. The epithelium of intestinal villi is mainly constituted of enterocytes and goblet cells (Kim and Ho 2010), while the outermost oral epithelium consists of a mitotically active basal cell layer, progressing through a number of differentiating intermediate layers to the superficial layers (Harris and Robinson 1992). It is currently believed that the superficial layers of the epithelium are the rate-limiting step of mucosal penetration (Squier and Rooney 1976). For example, in the oral cavity, based on the relative thickness of the epithelium and degree of keratinization, the drug permeability of the oral mucosa decreases in the order of sublingual > buccal > palatal (Petelin et al. 1998).

Mucus is a viscous colloid that forms a thin, continuous gel blanket adherent to the mucosal epithelial surface. The mean thickness of this layer varies from about 50–450 μm in humans (Sellers et al. 1986), which can be secreted by specialized mucus-secreting cells like the goblet cells or by special exocrine glands with mucous cells acini (Harris and Robinson 1992). Mucus contains antiseptic enzymes such as lysozymes, immunoglobulins, inorganic salts, proteins, and glycoproteins known as mucins (Kim and Ho 2010). Mucins consist of a protein core with branched oligosaccharide chains, and these mucin molecules are linear, flexible, and negatively charged due to the sialic acid and sulfate residues present on the mucin molecules (Boddupalli et al. 2010). Mucins are responsible for the gel-like properties of the mucus (Salcedo et al. 2012), which may play a role in cell–cell adhesion, as well as acting as a lubricant, allowing cells to move relative to one another (Harris and Robinson 1992).

The interaction between the mucus and carrier polymers is responsible for the formation of the adhesive interface (Bures et al. 2001). Mucoadhesives are synthetic or natural polymers that interact with the mucus layer covering the mucosal epithelial surface and mucin molecules. For mucoadhesion to occur, a succession of phenomena is required (Fig. 1.2). First, an intimate contact between a mucoadhesive and a membrane occur; second, after contact is established, penetration of the mucoadhesive into the crevice of the tissue surface or interpenetration of the chains of the mucoadhesive with those of the mucus takes place (Boddupalli et al. 2010). Several theories have been proposed to explain the fundamental mechanisms of adhesion (Longer and Robinson 1986; Mikos and Peppas 1990). These theories are related to

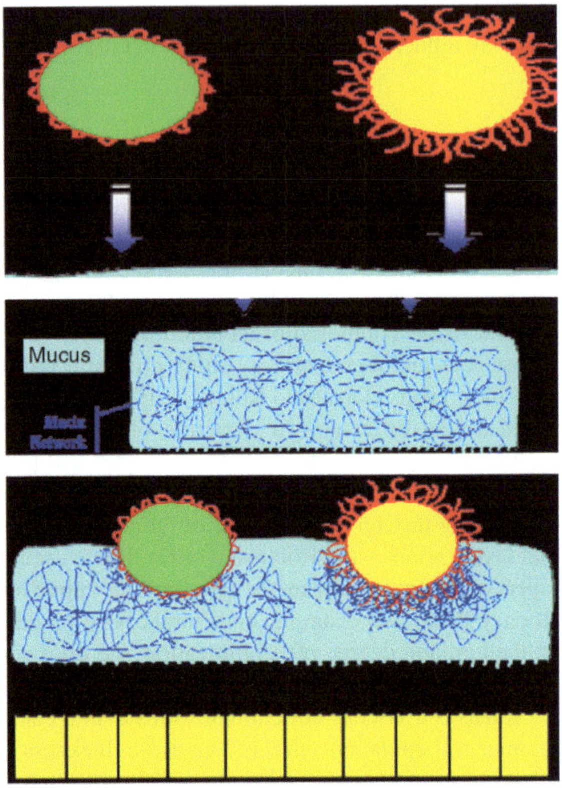

Fig. 1.2 Mucus interaction with drug delivery systems. Reprinted from reference Madhav et al. (2009), Copyright (2009), with permission from Elsevier

the formation of an electrical double layer at the interface due to the electron transfer between an adhesive polymer and a mucus glycoprotein network; the surface forces are acting between the groups in the two surfaces, including chemical bonds, electrostatic forces, van der Waals forces, and hydrogen and hydrophobic bonds, or the semi-permanent adhesive bond because of a sufficient depth diffusion of the polymer chains and the mucus mix.

Moreover, the specific mucosal surfaces can be targeted using site-specific chemical agents that are anchored onto the polymeric DDS to overcome the lack of specific binding to any mucosal surface. The development of carriers endowed with mucus or cell-specific ligands have increased therapeutic benefit and made site-specific trans-mucosal DDSs possible. Any ligand with a high binding affinity for mucin can be covalently linked to the carriers and be expected to influence the binding of carriers. For example, bacteria are able to adhere to epithelial surfaces of the enterocytes with the aid of fimbriae, which are long, lectin-like proteins found on the surface of many bacterial strains (Bernkop-Schnürch et al. 1995).

1.2.2 Interactions with Cells

The cell membrane is a bilayer of amphiphilic lipids with hydrophilic heads and hydrophobic tails (Edidin 2003), which can protect intracellular components from the surrounding environment, and controls the entry and exit of small molecules, nutrients, and particles (Brown et al. 1987; Shi and Massagué 2003). Small biomolecules can enter the cell by passive diffusion by their hydrophobicity/hydrophilicity properties, while in some cases, entry is regulated through other mechanisms (e.g., receptor or transporter). When carriers (mainly microparticles (MPs) and nanoparticles (NPs)) reach the exterior membrane of a cell, they can interact with components of the plasma membrane or extracellular matrix and enter the cell, mainly through endocytosis. Moreover, depending on the cell type, as well as the proteins, lipids, and other molecules involved in the process, endocytosis pathways can be classified into five types: clathrin-mediated endocytosis, caveolin-mediated endocytosis, clathrin/caveolae-independent endocytosis, micropinocytosis, and phagocytosis (Doherty and McMahon 2009; Kumari et al. 2010).

Except for phagocytosis, the former four mechanisms of those pathways are more common and occur in many cell types. Each pathway involves a unique set of membrane compositions and acts on particular types of particles. For example, clathrin-mediated endocytosis is the main mechanism of cell entry, which takes place in an area of the plasma membrane that is rich in clathrin (Brown and Petersen 1999). This unique protein is responsible for the spontaneous co-assembly formation of clathrin-coated pits, and with the aid of clathrin triskelions and accessory proteins, a hexagonal lattice is formed, inducing the invagination of the plasma membrane (Ungewickell and Branton 1981). While caveolae are flask-shaped membrane invaginations present in epithelial and non-epithelial cells and are lined with caveolin proteins, dimeric proteins are involved in the formation of their characteristic flask shape (Pelkmans and Helenius 2002). Many other proteins also play important roles in this process for structural stabilization, membrane curvature or vesicle fusion, such as cavin proteins, which are vesicle-associated membrane proteins (Nabi 2009). However, clathrin/caveolae-independent endocytosis takes place in cells requiring a specific lipid composition (mostly cholesterol) without clathrin and caveolae (Kumari et al. 2010). Moreover, micropinocytosis, which is important in the uptake of larger particles (Kuhn et al. 2014), does not involve the utilization of lipid rafts or pit-forming proteins and occurs when large membrane extensions caused by cytoskeleton rearrangement happen to form a large vesicle that traps particles (Norbury et al. 1995). Notably, compared with other three pathways, which frequently end up in degradative lysosomes (Doherty and McMahon 2009; Kumari et al. 2010; Norbury et al. 1995) and may not be suitable for carrier materials susceptible to degradation by lysosomal enzymes, the caveolae-dependent route can bypass lysosomes (Parton and Simons 2007; Xia et al. 2008). Most substances endocytosed by caveosomes are eventually delivered to the endoplasmic reticulum or the Golgi apparatus (Nomura 2005; Yousuf et al. 2013).

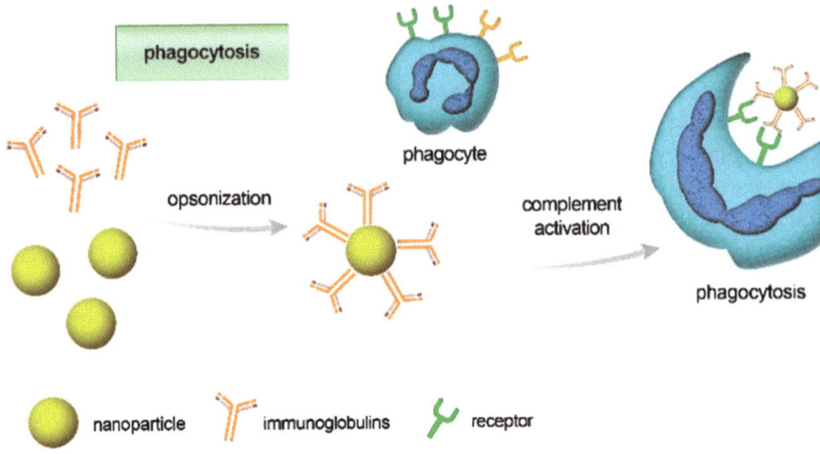

Fig. 1.3 Schematic illustration of the opsonization process. Reproduced from reference Behzadi et al. (2017), Copyright (2017), with permission from the Royal Society of Chemistry

Phagocytosis occurs primarily in professional phagocytes [e.g., macrophages, monocytes, neutrophils, and dendritic cells (DCs)], which are mainly responsible for host defense and the immune system in the human body (Aderem and Underhill 1999). The phagocytosis of NPs is usually initiated by opsonization, i.e., the adsorption of opsonins onto the surface of NPs (Fig. 1.3). Opsonins can be immunoglobulins, complement proteins, or other blood proteins (Aderem and Underhill 1999; Swanson 2008). Opsonized particles are subsequently identified through receptors on phagocytic cells and internalized. Phagocyte receptors involved in this process include Fc receptors, complement receptors, and other receptors such as mannose/fructose receptors and scavenger receptors (Behzadi et al. 2017).

Because of the immunostimulatory properties of biological therapeutics or NPs (Behzadi et al. 2017), this route of the pathway is closely related to the fate of NPs in human body involving in vivo distribution and tissue targeting efficiency. NPs internalized by this route are subject to be distributed in the RES organs and be cleared by them. RES, as a part of the immune system, includes monocytes and macrophages, which are located largely in the liver, spleen, lymph nodes, and bone marrow (Storm et al. 1995). As a result of recognition and phagocytosis by both macrophages in the blood and in these RESs, NPs are commonly sequestered to these organs and be cleared from the body by hepatic clearance (via the liver) or by the kidneys (Buxton and Benet 1996). Liver and spleen are two of the most extensively studied RES organs. Approximately, one-third of the liver macrophages are Kupffer cells (KCs), which line in the liver sinusoids, and are responsible for clearing large particles from the bloodstream (Bradfield 1974). Also, KCs are involved in the pathogenesis of liver injury through the release of biological cytokines (e.g., interleukin (IL) 6, tumor necrosis factor (TNF) α), superoxide, nitric oxide, eicosanoids, and chemokine (Racanelli and Rehermann 2006). In the spleen, macrophages exist in the red pulp trabeculae and venous sinuses where they can phagocytose and sequester the NPs. NPs highly susceptible to opsonization can preferentially accumulate in RES organs such as the liver and spleen. The rapid sequestration of intravenously injected colloidal NPs from the blood by RES cells, especially hepatic KCs, is problematic for the efficient targeting of drug carriers or diagnostic agents at a desired non-macrophage sites population. As a result, there has been a growing interest in the engineering of colloidal carrier systems that upon intravenous injection avoid rapid recognition by the RES and adequately remain in the blood.

Another characteristic of the immune system that has attracted scientists is its mucosal immune response, which can be initiated by the vaccines or antigens internalized by specialized antigen-sampling membranous (M) cells in the mucosa (Neutra et al. 1996). M cells are primarily located within the lymphoid tissue of mucosa in different parts of human body, such as follicle-associated epithelium in the intestinal mucosa (Neutra et al. 1996) and nasal-associated lymphoid tissue, which includes the tonsils and is known as Waldeyer's ring (Ermak et al. 1995). Take intestinal M cells as an example, after passage through the M cells, the particles/antigens are presented from the lumen to underlying lymphocytes and macrophages located intra- and subepithelially, triggering an immune response (Fig. 1.4). Although they are present in a smaller quantity than enterocytes, M cells display an enhanced transcytosis ability and reduced levels of membrane hydrolase activity, which favor the uptake of protein-decorated carriers or vaccines without degradation (Bye et al. 1984).

1.2.3 Interactions with Biomacromolecules

When carriers are administrated, their interactions with biomacromolecules can play a vital role in the internalization of cells and the performance and bioavailability of DDSs. Biomacromolecules, especially proteins, are essential parts of organisms and participate in every process within cells. Those biomacromolecules have structural or functional roles in the tissues, cells, and are important in the metabolic process of those tissues and cells. This part will focus on the interaction of carrier materials with proteins varying in the different route of administration to improve the absorption or targeting efficiency of DDSs, due to the important function of proteins in cell signaling, immune responses, and cell adhesion (Hannigan et al. 1996; Savill et al. 2002).

For drugs to pass through the mucosa, the intercellular tight junction is one of the major barriers to the paracellular transport of hydrophilic drugs. Tight junctions existing between cells are composed of transmembrane proteins (occludin and claudins) and junctional adhesion molecules (Tsukita et al. 2001). These proteins intercalate with cytoplasmic plaque proteins from adjacent cells to form an intercellular barrier including the membrane proteins zonula occludens (ZO) (including ZO-1, ZO-2, and ZO-3), and cingulin (Rescigno et al. 2001). ZO-1 and occludin phosphorylation were tested to be related to the stimulus-induced tight junction disassembly and paracellular permeability changes, suggesting these molecules could play a major

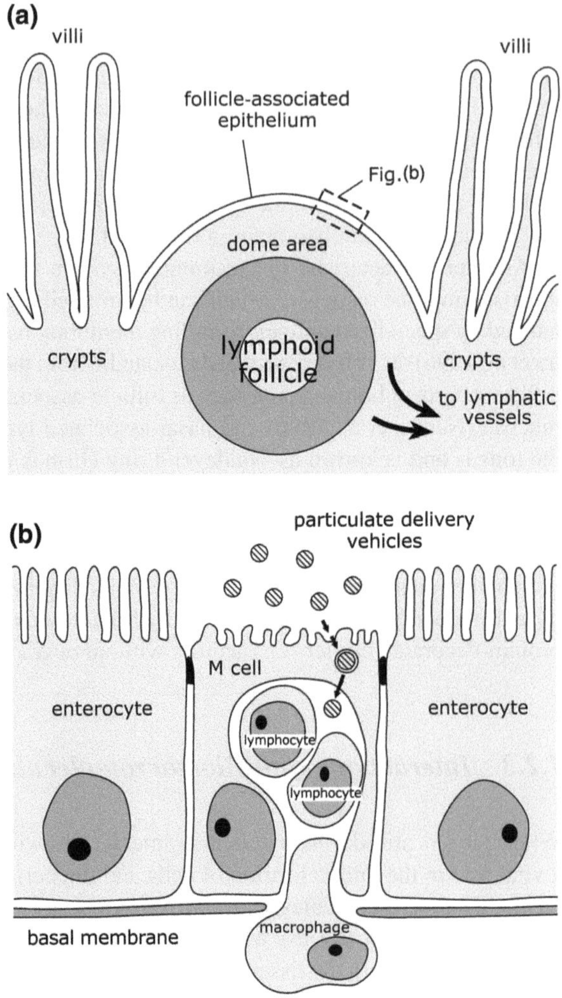

Fig. 1.4 Schematic transverse sections of a Peyer's patch lymphoid follicle and overlying follicle-associated epithelium (FAE), depicting membranous (M) cell transport of particulate delivery vehicles. The general structure of intestinal organized mucosa-associated lymphoid tissues is represented by the schematic transverse section of a Peyer's patch lymphoid follicle and associated structures in (a). The FAE is characterized by the presence of specialized antigen-sampling M cells (b). Reprinted from reference Clark et al. (2001), Copyright (2001), with permission from Elsevier

role in tight junction functional regulation (Reinhold and Rittner 2017). The ability of carrier polymers to disrupt tight junctions and open the paracellular route is usually based on the modulation of those proteins. Besides, it can also be achieved by many potential physicochemical factors, which regulate tight junction structure and permeability, including the concentration of cyclic adenosine monophosphate and intracellular calcium concentrations (Madhav et al. 2009).

When NPs enter the bloodstream for parenteral use, they immediately encounter a complex environment of plasma proteins and are prone to adsorb a significant amount of plasma proteins, including immunoglobulins and components of the complement system, as well as other abundant blood serum proteins such as fibrinogen and albumin (Göppert and Müller 2005). These proteins are thought to be important in the clearance process of NPs (Owens and Peppas 2006). Among these proteins, the interaction with the nanomaterial surface of opsonin proteins leads to a "protein corona" (Lundqvist et al. 2008; Monopoli et al. 2012). The newly formed protein corona can be divided into the "hard" and "soft" protein coronas, which differ in the binding strength and exchange rates of the proteins associated with a nanomaterial surface. A "hard" protein corona consists of proteins that are tightly bound with high binding affinity, which is also often referred to as irreversible protein binding (Milani et al. 2012). In contrast, proteins that are loosely bound to the nanomaterial surface, or are connected via weak protein–protein interactions, form a so-called "soft" protein corona (Lundqvist et al. 2008; Milani et al. 2012). At the molecular scale, it has been suggested that at the NP–cell membrane interface, the "hard" protein corona around the NP interacts with cell receptors and determines the route of particle internalization (Hu and Zhou 2013). The adsorption of proteins on the NPs surface, especially the "hard" protein corona, can increase its overall size, change its surface charge or facilitate its aggregation, causing a dramatic shift in the physicochemical properties of the NPs (Karmali and Simberg 2011; Lacerda et al. 2009).

References

Aderem A, Underhill DM (1999) Mechanisms of phagocytosis in macrophages. Annu Rev Immunol 17:593–623

Arora S, Ali J, Ahuja A, Khar RK, Baboota S (2005) Floating drug delivery systems: a review. AAPS PharmSciTech 6:E372–E390. https://doi.org/10.1208/pt060347

Bark DL, Ku DN (2010) Wall shear over high degree stenoses pertinent to atherothrombosis. J Biomech 43:2970–2977

Behzadi S et al (2017) Cellular uptake of nanoparticles: journey inside the cell. Chem Soc Rev 46:4218–4244. https://doi.org/10.1039/c6cs00636a

Bernhardt H, Knoke M (1997) Mycological aspects of gastrointestinal microflora. Scand J Gastroenterol 32:102–106. https://doi.org/10.1080/00365521.1997.11720731

Bernkop-Schnürch A (1998) The use of inhibitory agents to overcome the enzymatic barrier to perorally administered therapeutic peptides and proteins. J Controlled Release 52:1–16

Bernkop-Schnürch A, Gabor F, Szostak MP, Lubitz W (1995) An adhesive drug delivery system based on K99-fimbriae. Eur J Pharm Sci 3:293–299

Boddupalli BM, Mohammed ZNK, Nath RA, Banji D (2010) Mucoadhesive drug delivery system: an overview. J Adv Pharm Technol Res 1:381–387. https://doi.org/10.4103/0110-5558.76436

Bradfield JB (1974) Control of spillover: the importance of kupffer-cell function in clinical medicine. The Lancet 304:883–886. https://doi.org/10.1016/S0140-6736(74)91213-6

Brown CM, Petersen NO (1999) Free clathrin triskelions are required for the stability of clathrin-associated adaptor protein (AP-2) coated pit nucleation sites. Biochem Cell Biol 77:439–448

Brown D, Gluck S, Hartwig J (1987) Structure of the novel membrane-coating material in proton-secreting epithelial cells and identification as an H+ATPase. J Cell Biol 105:1637–1648

Bures P, Huang Y, Oral E, Peppas NA (2001) Surface modifications and molecular imprinting of polymers in medical and pharmaceutical applications. J Controlled Release 72:25–33

Buxton ILO, Benet LZ (1996) Pharmacokinetics: the dynamics of drug absorption, distribution, metabolism, and elimination. In: Brunton LL, Chabner BA, Knollmann BC (eds) Goodman & Gilman's: the pharmacological basis of therapeutics, 12th edn. pp 3–27

Bye W, Allan C, Trier J (1984) Structure, distribution, and origin of M cells in Peyer's patches of mouse ileum. Gastroenterology 86:789–801
Clark MA, Jepson MA, Hirst BH (2001) Exploiting M cells for drug and vaccine delivery. Adv Drug Deliv Rev 50:81–106. https://doi.org/10.1016/S0169-409X(01)00149-1
Cole AJ, David AE, Wang J, Galban CJ, Hill HL, Yang VC (2011) Polyethylene glycol modified, cross-linked starch-coated iron oxide nanoparticles for enhanced magnetic tumor targeting. Biomaterials 32:2183–2193. https://doi.org/10.1016/j.biomaterials.2010.11.040
Devasagayam T, Tilak J, Boloor K, Sane KS, Ghaskadbi SS, Lele R (2004) Free radicals and antioxidants in human health: current status and future prospects. J Assoc Physicians India Japi 52:794–804
Doherty GJ, McMahon HT (2009) Mechanisms of endocytosis. Annu Rev Biochem 78:857–902. https://doi.org/10.1146/annurev.biochem.78.081307.110540
Edidin M (2003) Lipids on the frontier: a century of cell-membrane bilayers. Nat Rev Mol Cell Biol 4:414. https://doi.org/10.1038/nrm1102
Ermak TH, Dougherty EP, Bhagat HR, Kabok Z, Pappo J (1995) Uptake and transport of copolymer biodegradable microspheres by Rabbit Peyer's patch M cells. Cell Tissue Res 279:433–436
Gao Z-G, Fain HD, Rapoport N (2005) Controlled and targeted tumor chemotherapy by micellar-encapsulated drug and ultrasound. J Controlled Release 102:203–222
Göppert TM, Müller RH (2005) Protein adsorption patterns on poloxamer-and poloxamine-stabilized solid lipid nanoparticles (SLN). Eur J Pharm Biopharm 60:361–372
Grabe M, Oster G (2001) Regulation of organelle acidity. J Gen Physiol 117:329–344. https://doi.org/10.1085/jgp.117.4.329
Guskiewicz KM, Mihalik JP (2011) Biomechanics of sport concussion: quest for the elusive injury threshold. Exerc Sport Sci Rev 39:4–11. https://doi.org/10.1097/JES.0b013e318201f53e
Han G, You CC, Kim BJ, Turingan RS, Forbes NS, Martin CT, Rotello VM (2006) Light-regulated release of DNA and its delivery to nuclei by means of photolabile gold nanoparticles. Angewandte Chemie 118:3237–3241
Hannigan GE et al (1996) Regulation of cell adhesion and anchorage-dependent growth by a new β1-integrin-linked protein kinase. Nature 379:91. https://doi.org/10.1038/379091a0
Harris D, Robinson JR (1992) Drug delivery via the mucous membranes of the oral cavity. J Pharm Sci 81:1–10
Hoffman AS (2013) Stimuli-responsive polymers: biomedical applications and challenges for clinical translation. Adv Drug Deliv Rev 65:10–16
Hu X, Zhou Q (2013) Health and ecosystem risks of graphene. Chem Rev 113:3815–3835
Jovic M, Sharma M, Rahajeng J, Caplan S (2010) The early endosome: a busy sorting station for proteins at the crossroads. Histol Histopathol 25:99–112
Karimi M et al (2016) Smart micro/nanoparticles in stimulus-responsive drug/gene delivery systems. Chem Soc Rev 45:1457–1501. https://doi.org/10.1039/c5cs00798d
Karmali PP, Simberg D (2011) Interactions of nanoparticles with plasma proteins: implication on clearance and toxicity of drug delivery systems. Expert Opin Drug Deliv 8:343–357
Kim YS, Ho SB (2010) Intestinal goblet cells and mucins in health and disease: recent insights and progress. Curr Gastroenterol Rep 12:319–330
Koçer A, Walko M, Meijberg W, Feringa BL (2005) A light-actuated nanovalve derived from a channel protein. Science 309:755–758
Kuhn DA, Vanhecke D, Michen B, Blank F, Gehr P, Petri-Fink A, Rothen-Rutishauser B (2014) Different endocytotic uptake mechanisms for nanoparticles in epithelial cells and macrophages Beilstein. J Nanotechnol 5:1625
Kumari S, Mg S, Mayor S (2010) Endocytosis unplugged: multiple ways to enter the cell. Cell Res 20:256. https://doi.org/10.1038/cr.2010.19
Lacerda SHDP, Park JJ, Meuse C, Pristinski D, Becker ML, Karim A, Douglas JF (2009) Interaction of gold nanoparticles with common human blood proteins. ACS Nano 4:365–379

Laurent S, Dutz S, Häfeli UO, Mahmoudi M (2011) Magnetic fluid hyperthermia: focus on superparamagnetic iron oxide nanoparticles. Adv Coll Interface Sci 166:8–23. https://doi.org/10.1016/j.cis.2011.04.003
Lee SH, Gupta MK, Bang JB, Bae H, Sung HJ (2013) Current progress in reactive oxygen species (ROS)-responsive materials for biomedical applications. Adv Healthc Mater 2:908–915
Likhitkar S, Bajpai AK (2014) An in vitro experimental approach to study magnetically targeted release of methotrexate from superparamagnetic starch nanocarriers. Int J Polym Mater Polym Biomater 63:941–950. https://doi.org/10.1080/00914037.2014.886232
Liu G et al (2003) Investigations of 99mTc morpholino pretargeting in mice. Nucl Med Commun 24:697–705
Longer M, Robinson J (1986) Fundamental-aspects of bioadhesion. Pharm Int 7:114–117
Lu Y, Sun W, Gu Z (2014) Stimuli-responsive nanomaterials for therapeutic protein delivery. J Controlled Release 194:1–19. https://doi.org/10.1016/j.jconrel.2014.08.015
Lundqvist M, Stigler J, Elia G, Lynch I, Cedervall T, Dawson KA (2008) Nanoparticle size and surface properties determine the protein corona with possible implications for biological impacts. Proc Natl Acad Sci 105:14265–14270
Madhav NS, Shakya AK, Shakya P, Singh K (2009) Orotransmucosal drug delivery systems: a review. J Controlled Release 140:2–11
Meng F, Cheng R, Deng C, Zhong Z (2012) Intracellular drug release nanosystems. Mater Today 15:436–442
Meyer DE, Shin B, Kong G, Dewhirst M, Chilkoti A (2001) Drug targeting using thermally responsive polymers and local hyperthermia. J Controlled Release 74:213–224
Mikos A, Peppas N (1990) Scaling concepts and molecular theories of adhesion of synthetic polymers to glycoproteinic networks. CRC Press, Boca Raton, FL
Milani S, Baldelli Bombelli F, Pitek AS, Dawson KA, Rädler J (2012) Reversible versus irreversible binding of transferrin to polystyrene nanoparticles: soft and hard corona. ACS Nano 6:2532–2541
Monopoli MP, Åberg C, Salvati A, Dawson KA (2012) Biomolecular coronas provide the biological identity of nanosized materials. Nat Nanotechnol 7:779–786
Mura S, Nicolas J, Couvreur P (2013) Stimuli-responsive nanocarriers for drug delivery. Nat Mater 12:991–1003
Murdan S (2003) Electro-responsive drug delivery from hydrogels. J Controlled Release 92:1–17
Nabi IR (2009) Cavin fever: regulating caveolae. Nat Cell Biol 11:789–791
Nesbitt WS et al (2009) A shear gradient–dependent platelet aggregation mechanism drives thrombus formation. Nat Med 15:665–673
Neutra MR, Frey A, Kraehenbuhl J-P (1996) Epithelial M cells: gateways for mucosal infection and immunization. Cell 86:345–348. https://doi.org/10.1016/S0092-8674(00)80106-3
Nomura R (2005) Caveolar endocytosis and virus entry. Uirusu 55:19–26
Norbury CC, Hewlett LJ, Prescott AR, Shastri N, Watts C (1995) Class I MHC presentation of exogenous soluble antigen via macropinocytosis in bone marrow macrophages. Immunity 3:783–791
Owens DE, Peppas NA (2006) Opsonization, biodistribution, and pharmacokinetics of polymeric nanoparticles. Int J Pharm 307:93–102. https://doi.org/10.1016/j.ijpharm.2005.10.010
Parton RG, Simons K (2007) The multiple faces of caveolae. Nat Rev Mol Cell Biol 8:185. https://doi.org/10.1038/nrm2122
Pelkmans L, Helenius A (2002) Endocytosis via caveolae. Traffic 3:311–320
Petelin M, Šentjurc M, Stolič Z, Skalerič U (1998) EPR study of mucoadhesive ointments for delivery of liposomes into the oral mucosa. Int J Pharm 173:193–202
Racanelli V, Rehermann B (2006) The liver as an immunological organ. Hepatology 43:S54–S62. https://doi.org/10.1002/hep.21060
Reinhold AK, Rittner HL (2017) Barrier function in the peripheral and central nervous system—a review. Pflug Arch Eur J Phy 469:123–134. https://doi.org/10.1007/s00424-016-1920-8
Rescigno M et al (2001) Dendritic cells express tight junction proteins and penetrate gut epithelial monolayers to sample bacteria. Nat Immunol 2:361–367

Rijcken C, Soga O, Hennink W, Van Nostrum C (2007) Triggered destabilisation of polymeric micelles and vesicles by changing polymers polarity: an attractive tool for drug delivery. J Controlled Release 120:131–148
Ritschel W (1991) Targeting in the gastrointestinal tract: new approaches. Methods Find Exp Clin Pharmacol 13:313–336
Saito G, Swanson JA, Lee K-D (2003) Drug delivery strategy utilizing conjugation via reversible disulfide linkages: role and site of cellular reducing activities. Adv Drug Deliv Rev 55:199–215
Salcedo I et al (2012) In vitro biocompatibility and mucoadhesion of montmorillonite chitosan nanocomposite: a new drug delivery. Appl Clay Sci 55:131–137. https://doi.org/10.1016/j.clay.2011.11.006
Sandra K (2009) Polysaccharides in oral drug delivery—recent applications and future perspectives. In: Edgar KJ, Heinze T, Buchanan CM (eds) Polysaccharide materials: performance by design, vol 1017. ACS symposium series, 1 edn. American Chemical Society, Washington DC, pp 13–30. https://doi.org/10.1021/bk-2009-1017.ch001
Savill J, Dransfield I, Gregory C, Haslett C (2002) A blast from the past: clearance of apoptotic cells regulates immune responses. Nat Rev Immunol 2:965–975
Scheline RR (1973) Metabolism of foreign compounds by gastrointestinal microorganisms. Pharmacol Rev 25:451–523
Sellers LA, Carroll NJH, Allen A (1986) Misoprostol-induced increases in adherent gastric mucus thickness and luminal mucus output. Dig Dis Sci 31:91S–95S. https://doi.org/10.1007/bf01309330
Shi Y, Massagué J (2003) Mechanisms of TGF-β signaling from cell membrane to the nucleus. Cell 113:685–700
Simon GL, Gorbach SL (1984) Intestinal flora in health and disease. Gastroenterology 86:174–193
Squier C, Rooney L (1976) The permeability of keratinized and nonkeratinized oral epithelium to lanthanum in vivo. J Ultrastruct Res 54:286–295
Storm G, Belliot SO, Daemen T, Lasic DD (1995) Surface modification of nanoparticles to oppose uptake by the mononuclear phagocyte system. Adv Drug Deliv Rev 17:31–48
Swanson JA (2008) Shaping cups into phagosomes and macropinosomes. Nat Rev Mol Cell Biol 9:639. https://doi.org/10.1038/nrm2447, https://www.nature.com/articles/nrm2447#supplementary-information
Torchilin VP (2000) Drug targeting. Eur J Pharm Sci 11:S81–S91. https://doi.org/10.1016/S0928-0987(00)00166-4
Tsukita S, Furuse M, Itoh M (2001) Multifunctional strands in tight junctions. Nat Rev Mol Cell Biol 2:285–293
Ungewickell E, Branton D (1981) Assembly units of clathrin coats. Nature 289:420. https://doi.org/10.1038/289420a0
Ventura A, Jacks T (2009) MicroRNAs and cancer: short RNAs go a long way. Cell 136:586–591
Wang W, Bao L, Lei J, Tu W, Ju H (2012) Visible light induced photoelectrochemical biosensing based on oxygen-sensitive quantum dots. Anal Chim Acta 744:33–38. https://doi.org/10.1016/j.aca.2012.07.025
Wang H et al (2013) Redox-responsive, core-cross-linked micelles capable of on-demand, concurrent drug release and structure disassembly. Biomacromolecules 14:3706–3712
Watts P, Smith A (2005) TARGIT technology: coated starch capsules for site-specific drug delivery into the lower gastrointestinal tract. Expert Opin Drug Deliv 2:159–167. https://doi.org/10.1517/17425247.2.1.159
Wootton DM, Ku DN (1999) Fluid mechanics of vascular systems, diseases, and thrombosis. Ann Rev Biomed Eng 1:299–329. https://doi.org/10.1146/annurev.bioeng.1.1.299
Xia T et al (2008) Comparison of the mechanism of toxicity of zinc oxide and cerium oxide nanoparticles based on dissolution and oxidative stress properties. ACS Nano 2:2121–2134
Yang L, Chu JS, Fix JA (2002) Colon-specific drug delivery: new approaches and in vitro/in vivo evaluation. Int J Pharm 235:1–15

Yang Y-H, Aloysius H, Inoyama D, Chen Y, Hu L-Q (2011) Enzyme-mediated hydrolytic activation of prodrugs. Acta Pharm Sin B 1:143–159

Yi Q, Sukhorukov GB (2014) UV light stimulated encapsulation and release by polyelectrolyte microcapsules. Adv Coll Interface Sci 207:280–289. https://doi.org/10.1016/j.cis.2013.11.009

Yousuf MA et al (2013) Caveolin-1 associated adenovirus entry into human corneal cells. PLOS ONE 8:e77462. https://doi.org/10.1371/journal.pone.0077462

Zhang Y, Yu J, Bomba HN, Zhu Y, Gu Z (2016) Mechanical force-triggered drug delivery. Chem Rev 116:12536–12563. https://doi.org/10.1021/acs.chemrev.6b00369

Chapter 2
Material Nature and Physicochemical Properties for High Performance of Carriers

Abstract As drugs are not inherently effective, the benefits of their treatment are closely related to the performance of drug delivery carriers. Given this, the interactions and stimulus responsiveness of the carrier materials within the human body play an important role in the performance of smart drug delivery systems (DDSs). Stimulated by advances in polymer materials, the stimulus responsiveness of drug carriers can be achieved by incorporating "to be triggered" constituting units or compositions within their own structure to acquire specific physicochemical properties. Thus, drugs can be released under specific stimuli resulting from the chemical degradation and physical conformation change of carriers. Another way to achieve stimulus responsiveness is to take advantage of drug–carrier interactions, whether they are chemical bonds or physical interactions. Once these drug–carrier interactions are broken or weakened, the drug will be released with responsiveness. On the other hand, the drug delivery efficiency of DDSs, especially target specificity, can be modulated by the physicochemical properties of carriers (e.g., hydrophobicity/hydrophilicity, size, shape, and surface charge) directing their interactions with tissues, cells, and biomacromolecules. By choosing appropriate materials of carriers and adjusting their structure and physicochemical properties, the well-controlled drug delivery and release of the therapeutics from carriers can be realized.

Keywords Fabrication materials strategy · Stimulus-responsive modality · Chemical degradation · Physical conformation · Particle size · Surface features

2.1 Material that Endows Stimulus Responsiveness

When being applied, stimulus-responsive carriers can undergo relatively abrupt physical or chemical changes in instant response to the applied stimulus. This lies in the fact that a specific stimulus can modify the structural (composition/conformation) or properties of the carriers, thereby promoting the release of the therapeutics to the specific biological environment (Fig. 2.1). The triggered release can be divided mainly into two major modes due to the different carrier strategies. When the biomolecule is entrapped within the carrier (Fig. 2.1), the release can be triggered by structural

J. Chen et al., *Drug Delivery Applications of Starch Biopolymer Derivatives*
https://doi.org/10.1007/978-981-13-3657-7_2

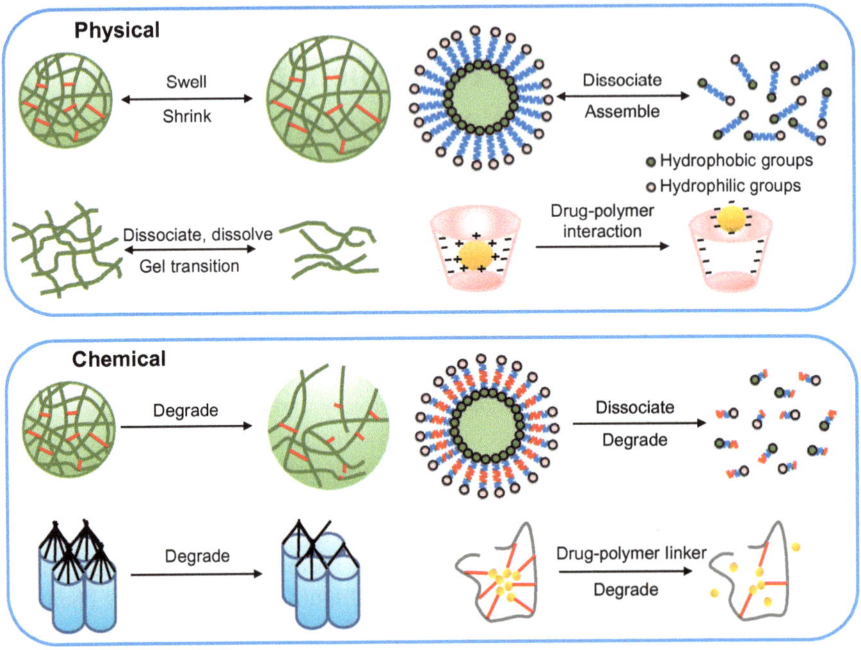

Fig. 2.1 Morphology (physical, chemical) changes of drug delivery carriers toward stimuli triggers

change within the carrier scaffold (i.e., the carrier degradation, the cleavage of the shell, and the change of aggregation conformation) (Fleige et al. 2012). While in some approaches (e.g., drug–polymer conjugates), the mechanism of release involves the breakage of the interaction (whether physical interaction or chemical linker) between the carrier and the bioactive agent (Fig. 2.1).

2.1.1 *Material Structure*

To achieve stimulus responsiveness, carriers need to be equipped with "to be triggered" constituting units or compositions within their structure. As a response, this functional part of the scaffold can lead to chemical degradation or physical deformation of the carrier itself, or changes of interactions between drug and carriers, thus facilitating the release of the payload near the target compartment. For example, carriers devised for enzymatic-triggered drug release incorporate either a structural scaffold that is susceptible to the corresponding enzyme degradation (e.g., degradable polymers) or a linker system between the drug and the carrier that works as an enzyme substrate (Fleige et al. 2012).

The design of pH-sensitive systems has generally been achieved by the incorporation of pH-tunable moieties within the structure of carriers. Ionizable chemical groups (e.g., amines and carboxylic acid) have been mostly chosen, as their hydrophobicity is altered by protonation and deprotonation (Foster et al. 2010; Liu et al. 2014). These functional groups endow degradation or the conformation changes of carriers along with the pH gradation. Especially, the protonation ability of polycations, such as poly(ethyleneimine)s (PEIs), is important for nanocarrier-mediated gene transfer approaches since it needs to act as proton sponges to escape the lysosomal destruction, which results in the release of polyplexes into the cytoplasm (Kichler 2004). Besides, for the conjugation of drugs or to render a polymeric backbone responsive, pH-sensitive linkages such as hydrazine (Kale and Torchilin 2007), hydrazide (Hrubý et al. 2005), and acetal (Gillies et al. 2004) are extensively used.

As mentioned earlier, the difference between oxidative and reductive environments can be used as a stimulus for controlled release. The design of reduction-sensitive systems has usually been achieved by using disulfide links. The disulfide bond, which is sensitive to intracellular GSH levels, is broken down into two sulfhydryl moieties and the drug release is triggered by the degraded carriers (Herlambang et al. 2011; Steinhilber et al. 2010). This feature makes thiol-containing structures within the carrier scaffold indispensable to design redox-responsive carriers.

For the response to an external force, the functionalized carriers generally involve more specialized functional units or compositions, such as a chromophore in the architecture to harvest the light for the fabrication of light-sensitive DDSs (Alvarez-Lorenzo et al. 2009). Moreover, electroresponsive gels typically contain repeat monomer units that are ionized at neutral pH and thus have a high number of a single charge spread out throughout the polymer backbone. Compared to polycationic and neutral polymers, polyanionic polymers have been mostly used, such as hyaluronic acid (HA), hydrolyzed polyacrylamide, and alginate (Murdan 2003). To acquire magnetic responsiveness, MNPs has been the most commonly used component to this end. However, these hard metal materials of MNPs are limited for magnetic force-triggered DDSs. Usually, the magnetic particles are combined with a flexible polymer material, with the advantage of not only stabilizing MNPs against aggregation and sedimentation but also improving their bioavailability and biocompatibility (Ding et al. 2014; Wang et al. 2011).

2.1.2 Physicochemical Properties

In the stimulus–response process, structural changes, especially conformation changes, can be closely related to the properties of carrier materials. The physicochemical properties have a huge impact on the responsive behavior of carriers in the case of some stimuli, such as deformation and dissociation. For example, the hydrophilicity/hydrophobicity balance of polymer micelles determines its self-assembly behavior and critical micelle concentration (CMC), which is a key parameter to evaluate the stability of micelles in the blood circulation system post-administration (Zhou et al. 2010). Moreover, as a response to many stimuli, physical deformation of carriers requires stretchable and flexible materials, such as hydrogel and elastomers, especially for mechanical-force-responsive DDSs (Wiggins et al. 2013). In the case of hydrogels, mechanical strength plays a vital role in the swelling–shrinkage behavior of hydrogels, as additional swelling can be resisted by an elastic polymeric network reaction force (Ismail et al. 2013). The suitable elasticity of hydrogel, which can be modulated by the physical or covalent cross-links within hydrogels, is required to inhibit its excessive swelling behavior, and simultaneously maintain its tensile strength in load-bearing applications.

For the response to internal or external stimuli, the properties of materials can also be used to endow the carriers with stimulus responsiveness, leading to the physical changes of the carriers thus facilitating the drug release. This can be exampled by the solubility of polymers in different biological mediums in human. Commonly used natural or synthetic polymers, such as pectin (Thakur et al. 1997), alginate (Haug and Larsen 1963), and various cellulose derivatives (Béchard et al. 1995; Schmidt and Niemann 1992), show pH-dependent solubility. Specifically, a dramatic increase in the solubility of pectin occurs in the acid medium, which may lead to the active ingredients due to the dissolution of carriers (Thakur et al. 1997). It should be mentioned that this feature of solubility can cause unintended drug release behavior, which needs to be overcome in DDSs. Thus, considerations should be taken for the choice of materials for the DDS design of a specific administration route. Another property that can be used or be overcome in a specific stimulus-responsive DDS is the degradability of carrier polymers. For the concern of biocompatibility and non-toxicity of drug delivery carriers, a certain degradability of materials is vital, which is the reason why natural polymers are popular in the application of DDSs. In addition, the enzymatic degradation of certain polymers and bonds can be utilized as a trigger of stimulus-responsive DDSs. However, as enzymes are widespread in the human body, to achieve triggered release at certain targeted sites, strategies have to be applied to avoid enzymatic degradation at the unwanted sites. For example, for microbial-enzyme-triggered colon-site-specific DDSs, the enzymatic degradation of carriers must be minimized in the upper GIT, and upon reaching the colon, the carrier polymers undergo assimilation by microorganisms and degradation by enzymes (Patten et al. 2009).

The phase transition of carrier polymers can be a trigger to the change of temperature to achieve responsiveness. The key feature for thermoresponsive carrier systems is the change of the hydration state, and therefore a change in the hydrophilic–hydrophobic balance depending on temperature. By changing the temperature around the upper critical solution temperature (UCST) or lower critical solution temperature (LCST), a phase transition leading to swelling or shrinking occurs (Fig. 2.2). A typical example of these polymers is poly(*N*-isopropylacrylamide) (PNIPAM), which undergoes a sol–gel transition at its LCST of 32 °C, and this value can be further optimized to be close to the body temperature through the introduction of hydrophobic groups (Yoshida et al. 1995). When the LCST of a polymer is between room temperature and body temperature, the polymer is endowed with inherent sensitivity toward the physiological temperature (Roy et al. 2013). Below the LCST, the composite carrier is in a swollen state, which is expected to cause faster diffusion of

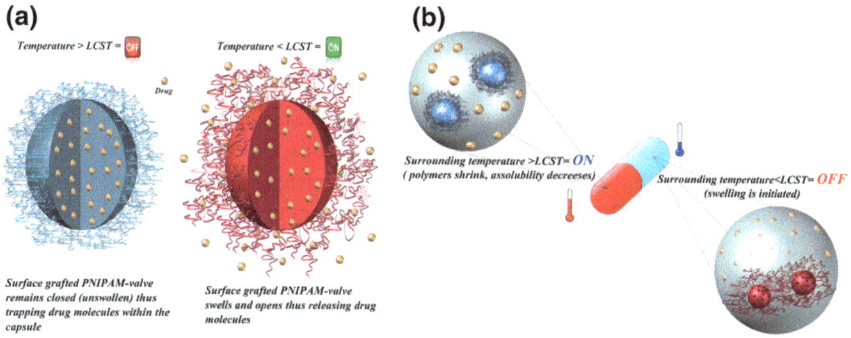

Fig. 2.2 Schematic illustration of on–off switch delivery of grafted poly(*N*-isopropylacrylamide) (PNIPAM) carriers. **a** Drug release in temperature below critical solution temperature (LCST). **b** Drug release in temperature above LCST. Adapted from Karimi et al. (2016) with permission from the Royal Society of Chemistry

the drug out from the matrix. Above the LCST, the carrier only slightly swells and the diffusion of drug molecules could be impaired.

2.2 Physicochemical Properties that Affect Interactions with Physiological Structures

It is now well understood that physicochemical properties of carriers (e.g., hydrophobicity/hydrophilicity, size, shape, surface charge, and elasticity) play a key role in directing their interactions with tissues, cells, and biomacromolecules. In particular, the interaction between NPs and cells is complicated, which can be affected by many factors. Modifying these factors can significantly reduce the immunotoxicity of NPs and make them useful platforms for drug delivery. In this section, the effects of these properties on the carrier interactions with physiological structures are focused and discussed as follows.

The roles of particle size and shape in cell uptake by endocytosis and of protein binding have also been investigated. In general, larger particles are taken up more efficiently than smaller particles of the same composition and surface properties (Fang et al. 2006). However, over a certain range of size, larger particles, despite multivalent receptor binding, are too large for the membrane wrapping that is necessary for efficient endocytosis (Johannes and Mayor 2010). Cancer cells are known to take up NPs of size between 50 and 150 nm with higher rates than healthy cells (Satchi-Fainaro et al. 2006). Moreover, the size of NPs can help particles to escape from the blood circulation to restricted sites, which are perturbed by inflammatory processes (e.g., rheumatoid arthritis, infarction, and infections) (Hallenbeck et al. 1986) or by tumors, although such defects are not a consistent feature of all tumors. Passive targeting toward the tumor cells can be achieved through an "enhanced permeability and retention" (EPR) effect, which is based on the mechanism that tumor tissues possess a "leaky" vasculature and impaired lymphatic drainage, allowing macromolecules and NPs to extravasate and accumulate more readily (Macda 2001; Maeda et al. 2000). Larger particles also experience more efficient uptake by phagocytes (Schäfer et al. 1992), which may induce the clearance by RES organs. For example, the size of the inter-endothelial cell slits of the spleen is approximately 200 nm in width, which means NPs larger than 200 nm are more likely to accumulate in the spleen (Moghimi et al. 2001). In addition to the size, the shape was also found to have a profound effect on the cellular uptake of NPs and protein binding. For instance, rod-shaped NPs have been demonstrated to undergo lower cellular uptake than spherical NPs (Chithrani and Chan 2007). The effect of shape may involve the membrane wrapping and the density of ligand binding on the surface of NPs during the cellular uptake process.

Surface features (especially surface charge) are the principal parameter dominating the NP interactions with cells, plasma proteins, and subsequent cell uptake, which are of great biological importance to the fate of NPs (Zhu et al. 2012). Evidence has been presented to illustrate the important role of electrostatic interactions in biological processes. For example, in blood circulation, NPs with negative surface charge were not hemolytic, and hemolytic tendency increased in proportion to the number of attached cationic surface groups (positive surface charge) (Dobrovolskaia et al. 2008). However, the biological interaction is a complicated process, and the effect of surface charge on the fate of NPs often varies with different polymeric particles and different interactions. One example that illustrates the complexity of this kind of interaction is the cell uptake of NPs. The charged NPs showed a more favorable thermo-dynamical interaction than their uncharged counterparts (Li and Gu 2010). Compared with the positively charged particles, there should be a slower uptake of negatively charged particles due to the negative "rejection" effect of the negatively charged cell membrane (Saikia et al. 2017). Nevertheless, the increased intensity of surface charge that favors cell uptake can also induce membrane disruption significantly. Anionic NPs induce less damage to plasma membrane integrity, mitochondrial and lysosomal compartments, and a higher extent of autophagosomes compared with cationic NPs (Lin et al. 2010). Accordingly, the charge density of particle surface cannot be too high to impair cell function and needs the consideration of the balance between the cellular uptake efficiency and the cell damage. Besides, particles bearing neutral surface charges have been shown to be less attractive to opsonization (Owens and Peppas 2006) and phagocytes (Zahr et al. 2006) than charged particles.

Concerning biocompatibility, a certain hydrophilic surface of carriers is highly recommended. Besides, evidence has been found that the adhesive and cohesive nature of the mucus gel was increased, when the water moved from the mucus gel to the contacting dry and partially hydrated mucoadhesive carriers (Mortazavi and Smart 1993). Regarding this, the hydrophilic content of carriers, which can affect the intermolecular forces responsible for the swelling of carriers, would be a significant factor in mucoadhesion. In addition, the surface hydrophobicity/hydrophilicity of NPs and interfacial forces play important roles in the interaction between NPs and cell membrane (Zhu et al. 2012). The hydrophobicity of an NP surface can change the protein corona around the NP, which may indirectly modify cell–NP interactions.

Specifically, hydrophobic particles are opsonized more quickly than hydrophilic particles (Owens and Peppas 2006). However, in the case of cell uptake process, due to the amphiphilic properties of membrane phospholipids, a "balanced" hydrophobicity/hydrophilicity is needed to enter the cell (Li et al. 2008).

The surface of particles can also be functionalized by the attachment of specific (but not macrophage recognizable) ligands onto the surface of carriers, which will open the possibility of targeting specific cell types or subsets of cells, i.e., active targeting delivery. This is based on the fact of the noteworthy biochemical differences between the vasculature of the tumor and the normal tissues. These ligands, which are usually upregulated on the surface of tumor-associated endothelial cells, include proteins involved in cell–cell or cell–matrix interactions (e.g., E-selectin, vascular cell adhesion molecule-1) and growth factors receptors (Koch et al. 1995; Todaro et al. 1980). Folic acid (FA), oligopeptides, and plasminogen (Avvakumova et al. 2014; Chau et al. 2004; Manoharan 2002) are among the most common markers for surface decoration of carriers to achieve active targeting.

The elasticity of particles refers to the ability of a material to resist deformation while under stress and to return to its original size/shape (Anselmo and Mitragotri 2017). Modulating particle elasticity is a viable strategy to control biological interactions, such as cellular binding and cellular internalization. Results showed that the NP elasticity may play a role in mediating phagocytosis by immune cells (Sosale et al. 2015). Despite the mixed results, generally, harder particles have been shown to be internalized to a higher extent than softer particles by macrophages and other immune cells (Anselmo and Mitragotri 2017). Even though determining how NP elasticity modulates interactions is essential, only recently has particle elasticity been studied to optimize particle–cell interactions. More work needs to be undertaken to acquire a general trend regarding soft versus hard particles for optimal cell interactions.

References

Alvarez-Lorenzo C, Bromberg L, Concheiro A (2009) Light-sensitive intelligent drug delivery systems. Photochem Photobiol 85:848–860

Anselmo AC, Mitragotri S (2017) Impact of particle elasticity on particle-based drug delivery systems. Adv Drug Deliv Rev 108:51–67. https://doi.org/10.1016/j.addr.2016.01.007

Avvakumova S et al (2014) Development of U11-functionalized gold nanoparticles for selective targeting of urokinase plasminogen activator receptor-positive breast cancer cells. Bioconjugate Chemistry 25:1381–1386

Béchard SR, Levy L, Clas S-D (1995) Thermal, mechanical and functional properties of cellulose acetate phthalate (CAP) coatings obtained from neutralized aqueous solutions. Int J Pharm 114:205–213

Chau Y, Tan FE, Langer R (2004) Synthesis and characterization of dextran–peptide–methotrexate conjugates for tumor targeting via mediation by matrix metalloproteinase II and matrix metalloproteinase IX. Bioconjugate Chemistry 15:931–941

Chithrani BD, Chan WC (2007) Elucidating the mechanism of cellular uptake and removal of protein-coated gold nanoparticles of different sizes and shapes. Nano Lett 7:1542–1550

Ding S, Shen Y, Walters KB, Chen J, Jin Y (2014) pH responsive behavior of Fe_3O_4@ PDEA-PEGMA core-shell hybrid magnetic nanoparticles. Int J Polym Mater Polym Biomater 63:487–492

Dobrovolskaia MA, Aggarwal P, Hall JB, McNeil SE (2008) Preclinical studies to understand nanoparticle interaction with the immune system and its potential effects on nanoparticle biodistribution. Mol Pharm 5:487–495

Fang C, Shi B, Pei Y-Y, Hong M-H, Wu J, Chen H-Z (2006) vivo tumor targeting of tumor necrosis factor-α-loaded stealth nanoparticles: effect of MePEG molecular weight and particle size. Eur J Pharm Sci 27:27–36

Fleige E, Quadir MA, Haag R (2012) Stimuli-responsive polymeric nanocarriers for the controlled transport of active compounds: concepts and applications. Adv Drug Deliv Rev 64:866–884. https://doi.org/10.1016/j.addr.2012.01.020

Foster S, Duvall CL, Crownover EF, Hoffman AS, Stayton PS (2010) Intracellular delivery of a protein antigen with an endosomal-releasing polymer enhances CD8 T-cell production and prophylactic vaccine efficacy. Bioconjugate Chemistry 21:2205–2212

Gillies ER, Goodwin AP, Fréchet JM (2004) Acetals as pH-sensitive linkages for drug delivery. Bioconjug Chem 15:1254–1263

Hallenbeck JM et al (1986) Polymorphonuclear leukocyte accumulation in brain regions with low blood flow during the early postischemic period. Stroke 17:246–253. https://doi.org/10.1161/01.str.17.2.246

Haug A, Larsen B (1963) The solubility of alginate at low pH. Acta Chem Scand 17:1653–1662

Herlambang S et al (2011) Disulfide crosslinked polyion complex micelles encapsulating dendrimer phthalocyanine directed to improved efficiency of photodynamic therapy. J Controlled Release 155:449–457

Hrubý M, Koňák Č, Ulbrich K (2005) Polymeric micellar pH-sensitive drug delivery system for doxorubicin. J Controlled Release 103:137–148

Ismail H, Irani M, Ahmad Z (2013) Starch-based hydrogels: present status and applications. Int J Polym Mater Polym Biomater 62:411–420. https://doi.org/10.1080/00914037.2012.719141

Johannes L, Mayor S (2010) Induced domain formation in endocytic invagination, lipid sorting, and scission. Cell 142:507–510

Kale AA, Torchilin VP (2007) Design, synthesis, and characterization of pH-sensitive PEG–PE conjugates for stimuli-sensitive pharmaceutical nanocarriers: the effect of substitutes at the hydrazone linkage on the pH stability of PEG–PE conjugates. Bioconjugate Chemistry 18:363–370

Karimi M et al (2016) Smart micro/nanoparticles in stimulus-responsive drug/gene delivery systems. Chem Soc Rev 45:1457–1501. https://doi.org/10.1039/c5cs00798d

Kichler A (2004) Gene transfer with modified polyethylenimines. J Gene Med 6:S3–S10

Koch AE, Halloran MM, Haskell CJ, Shah MR, Polverini PJ (1995) Angiogenesis mediated by soluble forms of E-selectin and vascular cell adhesion molecule-1. Nature 376:517. https://doi.org/10.1038/376517a0

Li Y, Gu N (2010) Thermodynamics of charged nanoparticle adsorption on charge-neutral membranes: a simulation study. J Phys Chem B 114:2749–2754

Li Y, Chen X, Gu N (2008) Computational investigation of interaction between nanoparticles and membranes: hydrophobic/hydrophilic effect. J Phys Chem B 112:16647–16653

Lin J, Zhang H, Chen Z, Zheng Y (2010) Penetration of lipid membranes by gold nanoparticles: insights into cellular uptake, cytotoxicity, and their relationship. ACS Nano 4:5421–5429

Liu J, Huang Y, Kumar A, Tan A, Jin S, Mozhi A, Liang X-J (2014) pH-sensitive nano-systems for drug delivery in cancer therapy. Biotechnol Adv 32:693–710

Maeda H (2001) The enhanced permeability and retention (EPR) effect in tumor vasculature: the key role of tumor-selective macromolecular drug targeting. Adv Enzyme Regul 41:189–207. https://doi.org/10.1016/S0065-2571(00)00013-3

Maeda H, Wu J, Sawa T, Matsumura Y, Hori K (2000) Tumor vascular permeability and the EPR effect in macromolecular therapeutics: a review. J Controlled Release 65:271–284

Manoharan M (2002) Oligonucleotide conjugates as potential antisense drugs with improved uptake, biodistribution, targeted delivery, and mechanism of action. Antisense Nucleic Acid Drug Develop 12:103–128

Moghimi SM, Hunter AC, Murray JC (2001) Long-circulating and target-specific nanoparticles: theory to practice. Pharmacol Rev 53:283–318

Mortazavi SA, Smart JD (1993) An investigation into the role of water movement and mucus gel dehydration in mucoadhesion. J Controlled Release 25:197–203

Murdan S (2003) Electro-responsive drug delivery from hydrogels. J Controlled Release 92:1–17

Owens DE, Peppas NA (2006) Opsonization, biodistribution, and pharmacokinetics of polymeric nanoparticles. Int J Pharm 307:93–102. https://doi.org/10.1016/j.ijpharm.2005.10.010

Patten GS, Augustin MA, Sanguansri L, Head RJ, Abeywardena MY (2009) Site specific delivery of microencapsulated fish oil to the gastrointestinal tract of the rat. Dig Dis Sci 54:511–521. https://doi.org/10.1007/s10620-008-0379-7

Roy D, Brooks WL, Sumerlin BS (2013) New directions in thermoresponsive polymers. Chem Soc Rev 42:7214–7243

Saikia C, Das MK, Ramteke A, Maji TK (2017) Evaluation of folic acid tagged aminated starch/ZnO coated iron oxide nanoparticles as targeted curcumin delivery system. Carbohyd Polym 157:391–399. https://doi.org/10.1016/j.carbpol.2016.09.087

Satchi-Fainaro R, Duncan R, Barnes CM (2006) Polymer Therapeutics for cancer: current status and future challenges. In: Satchi-Fainaro R, Duncan R (eds) Polymer Therapeutics II, 1st edn. Springer, Berlin, Heidelberg, pp 1–65. https://doi.org/10.1007/12_024

Schäfer V et al (1992) Phagocytosis of nanoparticles by human immunodeficiency virus (HIV)-infected macrophages: a possibility for antiviral drug targeting. Pharmaceutical Research 9:541–546

Schmidt PC, Niemann F (1992) The MiniWiD-coater: II. Comparison of acid resistance of enteric-coated bisacodyl pellets coated with different polymers. Drug Develop Ind Pharm 18:1969–1979

Sosale NG, Rouhiparkouhi T, Bradshaw AM, Dimova R, Lipowsky R, Discher DE (2015) Cell rigidity and shape override CD47's "self"-signaling in phagocytosis by hyperactivating myosin-II. Blood 125:542–552. https://doi.org/10.1182/blood-2014-06-585299

Steinhilber D, Sisson AL, Mangoldt D, Welker P, Licha K, Haag R (2010) Synthesis, reductive cleavage, and cellular interaction studies of biodegradable, polyglycerol nanogels. Adv Funct Mater 20:4133–4138

Thakur BR, Singh RK, Handa AK, Rao M (1997) Chemistry and uses of pectin—a review. Crit Rev Food Sci Nutr 37:47–73

Todaro GJ, Fryling C, De Larco JE (1980) Transforming growth factors produced by certain human tumor cells: polypeptides that interact with epidermal growth factor receptors. Proc Natl Acad Sci 77:5258–5262

Wang S, Zhou Y, Niu H, Zhang X (2011) Layer-by-Layer self-assembly of polyaspartate and Poly (ethyleneimine) on magnetic nanoparticles: Characterization and adsorption of protein. Current Applied Physics 11:1337–1342

Wiggins KM, Brantley JN, Bielawski CW (2013) Methods for activating and characterizing mechanically responsive polymers. Chem Soc Rev 42:7130–7147

Yoshida R, Uchida K, Kaneko Y, Sakai K, Kikuchi A, Sakurai Y, Okano T (1995) Comb-type grafted hydrogels with rapid deswelling response to temperature changes. Nature 374:240–242. https://doi.org/10.1038/374240a0

Zahr AS, Davis CA, Pishko MV (2006) Macrophage uptake of core–shell nanoparticles surface modified with poly (ethylene glycol). Langmuir 22:8178–8185

Zhou H, Yu W, Guo X, Liu X, Li N, Zhang Y, Ma X (2010) Synthesis and characterization of amphiphilic glycidol–chitosan–deoxycholic acid nanoparticles as a drug carrier for doxorubicin. Biomacromolecules 11:3480–3486

Zhu M, Nie G, Meng H, Xia T, Nel A, Zhao Y (2012) Physicochemical properties determine nanomaterial cellular uptake, transport, and fate. Acc Chem Res 46:622–631

Chapter 3
Starch

Abstract Among all polysaccharides, starch has been widely used in the pharmaceutical domain due to its low cost, its various sources, and high availability. Starch is organized in discrete granules whose morphology and supramolecular structure depend on the organization's way of amylose and amylopectin. The granule architecture determines the accessibility of starch to water or enzymes, thus affecting the gelatinization and enzymatic hydrolysis behavior. Gelatinization of starch is vital for its functional properties; and once on cooling, the gelatinized starch undergoes retrogradation, with the starch molecules reassociating into partially ordered structures that differ from those in native granules. These structural changes of starch may cause significant differences in its digestion. Generally, starch can be digested by human enzymes, which gives starch great biodegradability. Yet, the fast degradation of starch could be a hurdle in the application of starch in drug delivery systems (DDSs). Therefore, natural starches should be modified by physical, chemical, and enzymatic methods in order to expand the application of starch as drug carriers. Armed with the structural, physicochemical, and enzymatic information of starch, the advanced and well-controlled starch-based DDSs can be realized.

Keywords Starch · Supramolecular structure · Gelatinization and retrogradation · Enzymatic hydrolysis · Resistant starch · Modified starch

3.1 Basic Characteristics of Starch

Starch is the second most abundant biomass material in nature, which is found in plant roots, stalks, crop seeds, and staple crops such as rice, corn, wheat, tapioca, and potato (Buléon et al. 1998). Starch occurs naturally as insoluble, semicrystalline granules, made up of two polymers of D-glucose: the essentially unbranched amylose and the highly branched amylopectin. Amylose is defined as linear molecules of anhydrous glucose units that are linked mainly by α-(1–4)-D-glycoside bonds, with the average molecular weight (M) less than a million (Buléon et al. 1998; Mua and Jackson 1997). Amylopectin, however, extensively branched with α-(1–6) linkages, has an average M up to hundreds of millions (Buléon et al. 1998; Mua and Jackson 1997).

J. Chen et al., *Drug Delivery Applications of Starch Biopolymer Derivatives*
https://doi.org/10.1007/978-981-13-3657-7_3

Most native starches contain 20–30% amylose, although natural variants outside this range occur widely (Englyst et al. 1992). Differences in the amounts of these two molecules and the way they organized within the granule give rise to considerable variability in the size, shape, and properties of starch granules between and within species.

Starch granules are usually characterized at several size scales (Fig. 3.1). The starch molecules are biosynthesized and densely packed in granules with dimensions ranging from 1 to 100 μm (Buléon et al. 1998; Tester et al. 2004). Starch granules are considered to have an amorphous core of mostly amylose (in non-waxy starches), which is surrounded by a concentric pattern of alternating semicrystalline and amorphous growth rings (Wang and Copeland 2013). The semicrystalline growth rings are made up of blocklets, which consist of amylopectin and amylose arranged into crystalline and amorphous lamellae with a repeat period of 9–11 nm; this spacing does not seem to vary with the botanical source of the starch (Gallant et al. 1997; Pérez and Bertoft 2010). The size and organization of the blocklets are considered to play a role in the resistance of starch to gelatinization and digestion (Gallant et al. 1997; Pérez and Bertoft 2010). The natural polydispersity of starch granules is reflected in variability in functional properties such as water absorption, swelling, pasting and gelling behavior, and susceptibility to enzyme attack.

3.2 Gelatinization and Retrogradation

Starch is highly hydrophilic and shows a high degree of swelling when put in contact with an aqueous medium. When heated in excess water, the starch granules undergo an irreversible phase transition, referred to as gelatinization, in which the highly ordered structure of starch is disrupted by water. Starch gelatinization has been broadly defined as "the collapse (disruption) of molecular orders (breaking of hydrogen bonds) within the starch granules manifested in irreversible changes in properties such as water uptake, granular swelling, crystallite melting, unwinding of double helices, loss of birefringence, starch solubilization and viscosity development" (Atwell et al. 1988; Biliaderis 2009). On heating, water first enters the amorphous regions, which expands and transmits disruptive forces into the crystalline regions (BeMiller 2011). These changes are accompanied by the swelling of the granules, which results in an increase in viscosity under mixing conditions before the eventual collapse of the granules to form a paste if the water content of the system is high enough. After gelatinization, the disrupted and leached-out starch polymers form a three-dimensional network that turns into a gel upon cooling. This process, called retrogradation, involves amylose reassociation in the first stages and amylopectin reassociation over a longer time-scale and to a lesser extent. The cohesiveness of retrograded gel is provided by the double helices formed by associated glycan chains (junction zones), and these associations can reach a further extent until partially recreated crystallinity, thus, increased enzymatic resistance (Haralampu 2000).

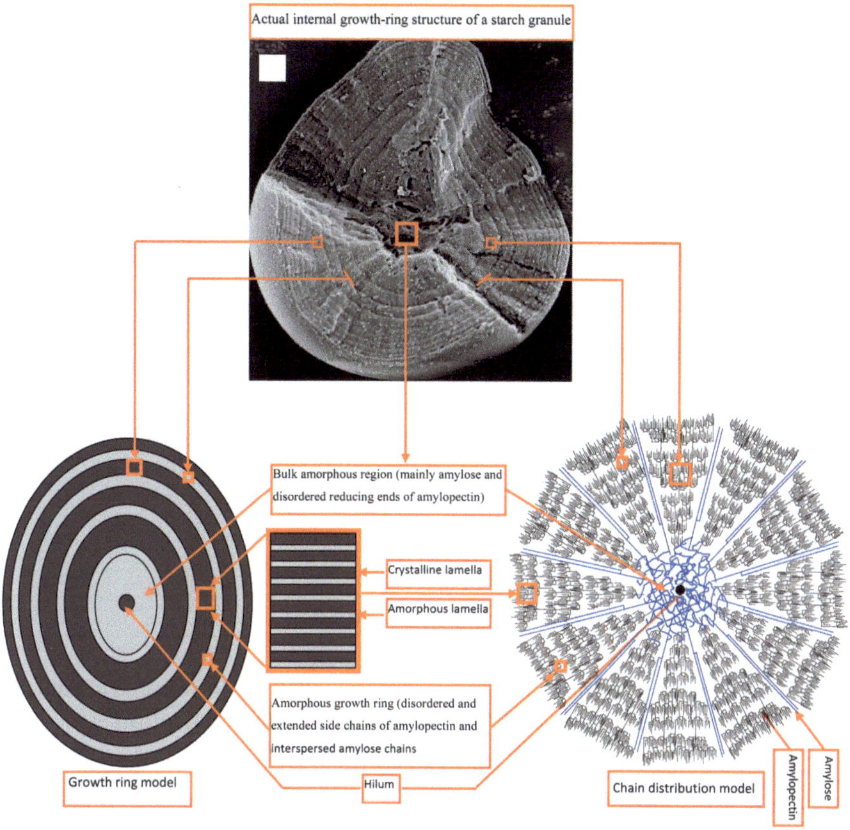

Fig. 3.1 Starch granule and its hierarchical structure. Adapted from Wang et al. (2012), copyright (2012), with permission from Elsevier

3.3 Digestion of Starch

Starch is a biodegradable polymer, which can be degraded very fast with the help of enzymes. The main starch-degradable enzyme in human is α-amylase, with its presence in saliva and pancreatic juice (Ramasubbu et al. 1996), and to a smaller extent in blood serum (Smith and Roe 1949). Some minor α-amylase activity is also present in other tissues including normal lung, adipose tissue, and skeletal muscle (Benedetti et al. 2004). In addition, there is some documentation of hyperamylasemia in tumors or nonmalignant pulmonary disorders (Benedetti et al. 2004; Sudo and Kanno 1976). The hydrolysis of α-amylase, which acts upon 1–4 glycan bonds of starch, results in a subsequent reduction in their M and hence, the loss of mechanical strength of starch granules. Those starch fractions which are incompletely digested and undigested by enzymes within a certain time has raised lots of interests by scientists (Cummings and Englyst 1991; Englyst et al. 1992). The term "resistant starch" (RS) is described as the fraction of starch that was resistant to enzymatic hydrolysis in the simulated intestinal medium after 120 min (Englyst et al. 1992). Moreover, because starch can be fermented by gut microflora, RS is now defined as the fraction of dietary fiber, which escapes digestion in the small intestine but can be degraded by the colon microorganisms (Thompson 2000). It is subdivided into five fractions: RS1, RS2, RS3, RS4 and RS5 (Sanz et al. 2009). RS1 represents physically inaccessible starch that is locked within cell walls and food matrixes, thus preventing amylolysis. RS2 is composed of native starch granules from certain plants containing uncooked starch that was gelatinized poorly and hydrolyzed slowly by α-amylases. RS3 is mainly retrograded amylose formed during cooling of gelatinized starch. RS4 is the RS where novel chemical bonds [other than α-(1–4) or α-(1–6)] are formed. While RS5 is amylose-lipid complex which is resistant to amylolytic enzyme hydrolysis (Holm et al. 1983). Modified starches obtained by various types of chemical treatments are included in this category. The modifications of starch to prepare RS provide a viable approach to achieve drug release at the targeted site in a more controlled and sustained way.

3.4 Starch Derivatives (Modified Starch)

Native starch may not be appropriate in DDSs, as many drugs are released too quickly from unmodified starch-based systems (Henrist et al. 1999; Michailova et al. 2001; Pereswetoff-Morath 1998). This is due to the substantial swelling and rapid enzymatic degradation of native starch. The fast degradation of starch could be a hurdle in the application of starch in DDSs. However, it is well known that α-amylase is less effective in degrading modified starch because of its molecular specificity (Wepner et al. 1999). For the application of starch in DDSs, various chemical, physical, and enzymatic modification methods could be used to improve the stability and reduce the degradation of starch (Table 3.1). Due to the presence of various derivatizable groups in molecular chains, starch can be easily modified, resulting in different enzymatic resistance, and various starch-based functional materials.

Modification of starch is an ongoing process as there are numerous possibilities to accentuate their positive characteristics, diminish their undesirable properties, and/or add new attributes. Chemical modifications, which are the most common industrial means of enhancing starch properties, entail reactions of starch with chemical reagents to introduce new chemical substituent groups (Wurzburg 1986). Chemical modifications are generally achieved through derivatization such as etherification, esterification, crosslinking, oxidation, cationization, and grafting of starch. The extent of modifications is most commonly quantified by the degree of substitution (DS) or molar degree of substitution (MS), crosslinking degree or oxidized degree. Besides chemical modifications, physical or enzymatic modifications have been investigated and developed rapidly in recent years. Physical methods such as hydrothermal, microwave, ionization radiation, sonication, ball milling, and extrusion have been applied to modify starch (Duan et al. 2007).

Table 3.1 Different starch derivatives used as the excipient of drug delivery system

Modified starch	Major features [Refs]	Application
Chemically modified starch		
Carboxymethyl starch (CMS)	Bringing anionic carboxyl groups of CMS to increase hydrophilicity Degree of substitution (DS) plays a more important role than amylose content; increased DS leads to delay release in acid medium (Kittipongpatana et al. 2009; Saboktakin et al. 2009; Tatongjai and Lumdubwong 2010)	Used for all dosage forms, mostly for pH-sensitive carriers
Hydroxyethyl starch (HES)	Nonionic starch derivative; bringing hydroxyl groups to starch molecules to increase hydrophilicity Better thermal stability and solubility HES: volume expanders, lacking immunogenicity; mostly used average molecular weights and DS of HES is 130 kDa/0.4 (Goszczynski et al. 2014; Li et al. 2014) and 200 kDa/0.5 (Fichter et al. 2013; Li et al. 2016)	Used for all dosage forms, especially for starch-drug conjugates
Starch esters	Esterification of starch with stearic acid, palmitoyl chloride, acetic anhydride, poly(ethylene glycol), and so on (Li et al. 2015a; Tan et al. 2010; Zhang et al. 2013a)	Used for forming micelles
Starch acetate (SA)	Raised hydrophobicity owing to the acetyl groups of SA More moisture resistant and thermoplasticity; overcoming the substantial swelling and rapid enzymatic degradation in biological fluids (Bie et al. 2016; Situ et al. 2015; Tuovinen et al. 2004a, b; Xiao et al. 2016) Mostly used: DS 1.9 or DS 2.6 (Singh et al. 2011)	Mostly used for tablets, films, microparticles (MPs), or nanoparticles (NPs); not suitable for hydrogel preparation
Propyl starch	Raised hydrophobicity Nature of the hydrophobic chain, as well as DS, facilitates the dissolution in low-risk organic solvents (Dandekar et al. 2012; Simi and Emilia Abraham 2007) Increasing DS favors the sustained release of hydrophobic drugs, e.g., DS 1.45 of propyl starch controls the release of hydrophobic drugs better than DS of 1.05 (Santander-Ortega et al. 2010)	Mostly used for tablets, films, MPs, or NPs; not suitable for hydrogel preparation

(continued)

Table 3.1 (continued)

Modified starch	Major features [Refs]	Application
Crosslinked starch	More hydrophobic nature compared with native starch (Kim et al. 2015) Crosslinkers: glutaraldehyde, epichlorohydrin, citric acid, phosphoric trichloride, sodium trimetaphosphate (Huang et al. 2013; Li et al. 2015b; Saikia et al. 2016; Tuma et al. 1982) Increased gel strength and water absorbance; decreased swelling capacity; resistance to shear; decreased gel cohesiveness, clarity, and water solubility, higher thermal stability	Used for all dosage forms, mostly used modification method
Grafted starch	Grafting copolymerization of hydrophilic vinyl monomers (Casas et al. 2009; Elvira et al. 2002; Kumar et al. 2015; Mahkam 2010; Saboktakin et al. 2009; Sadeghi and Hosseinzadeh 2008; Shaikh et al. 2015) Vinyl monomers: acrylic acid, acrylamide, acrylonitrile, methacrylate, and its derivative Initiator: benzoyl peroxide, 4-dimethylaminobenzyl alcohol, potassium persulfate, ceric ammonium nitrate, ammonium persulfate; *N, N, N′, N′*-tetramethylethylenediamine; crosslinker: *N, N′*-methylene bisacrylamide Better thermal stability; higher hydrodynamic radius and hydrodynamic volume; stronger water absorbency; greater bioadhesion Enlarged particle size and decreased density and moisture content; more amorphous than raw starch, higher glass transition temperature Graft with other monomers (Balmayor et al. 2009; Jiang et al. 2003; Zhang et al. 2013b) Other monomers: L-glutamic acid, ε-caprolactone, L-lactide, and so on Grafting reaction prevents the erosion of enzymes, and results in more hydrophobic materials	Used for all dosage forms, mostly used for hydrogel preparation
Oxidized starch	Oxidizers: 2,2,6,6-tetramethyl-1-piperidi-nyloxy, sodium hypochlorite, ceric ammonium nitrate, hydrogen peroxide, persulfate (Kuakpetoon and Wang 2001; Sánchez-Rivera et al. 2005; Thiele et al. 2017; Yamada et al. 2014) Primarily oxidation occurs at the hydroxyl groups of C-2, C-3, and C-6 positions on a D-glucopyranosyl unit, oxidizing hydroxyl groups to carbonyl groups, and then to carboxyl groups Higher whiteness, solubility; better water absorbency; lower pasting viscosity, gelatinization temperature, and enthalpy values; reduced gelatinization and retrogradation tendencies; increased in gel hardness (Chan et al. 2011; Oladebeye et al. 2013; Vanier et al. 2012) Dialdehyde starch is an important oxidized starch, of which the thermal stability and mechanical properties increased (Zhang et al. 2007)	Mostly used for hydrogel, MPs, NPs, in some cases, used for further modification

(continued)

Table 3.1 (continued)

Modified starch	Major features [Refs]	Application
Cationic starch	Cationic reagents containing the amino, imino, ammonium, or sulfonium groups (Constantin et al. 2003; Fonseca et al. 2015; Sieradzki et al. 2014) Increasing DS to 0.07 to reach cold water solubility Higher pasting viscosity and lower gelatinization temperature with increasing DS; higher solubility and stability; good flowability	Combining with anionic polymers, or used for gene delivery
Acid-hydrolyzed starch	Acidifiers: organic acids, HCl, H_2SO_4 (da Rosa Zavareze et al. 2012; Wuttisela et al. 2009) Reduced granule size and increased crystallinity; loss in pasting viscosities; increased water solubility Broader range of gelatinization temperature, and decreased tendency of retrogradation	Used for starch crystalline NPs
Blend with other polymers	Blend with synthetic degradable polymers (Lu et al. 2009; Namazi and Belali 2009, 2016) Polyesters: polyvinyl alcohol, polylactide, Carbopol 974P Blend with biopolymers (Desai 2005; Lefnaoui and Moulai-Mostefa 2011, 2014; Oechslein et al. 1996; Sivapragasam et al. 2014; Soares et al. 2013; Wang et al. 2010; Xiao and Fang 2009; Xu et al. 2014) Natural polymers such as chitosan, pectin, alginate, carboxymethyl cellulose, kappa carrageenan Improve mechanical properties, water barrier properties, miscibility and thermostability; improved biodegradability and mucoadhesive properties	Used for all dosage forms to enhance multifunctional properties
Physically modified starch		
Retrograded starch (RS)	Recrystallization, thus raises the enzymatic resistance (Marsh and Blanshard 1988) The polymorph of RS: more B-type crystals characterized by double helices (Yoon et al. 2009) Optimal condition: retrograded and recrystallized at 4 °C and 50–60% of water content; mild heating during refrigeration facilitates the growth of starch crystals (Tako 1996)	Mostly used for all dosage forms and for colon-targeted drug delivery
Enzymatically modified starch	Molecular weight distributions of starch macromolecules together with the chain lengths affect the properties such as viscosity, texture, and enzymatic resistance (Castro et al. 2005; Situ et al. 2014)	Mostly used for modification of the molecular characterization of starch

Enzymatic modifications mainly use hydrolyzing enzymes, such as pullulanase and isoamylase, to enzymatically hydrolyze the α-1,6 glycosidic bonds of amylopectin selectively (Qiao et al. 2006). The digestibility of debranched products is related to the molecular and crystallized structure as well as morphology during the debranching and recrystallization of native starch (Cai and Shi 2010; Cai et al. 2010). However, in some cases, one single modification method cannot meet the requirements of DDSs, thus dual or multiple modifications have been investigated, such as crosslinking/blending (Larionov et al. 1999), crosslinking/oxidation (Li et al. 2011), and high-pressure/high-temperature/enzymatic modification (Situ et al. 2014). Those multi-modifications can bring multifunctional properties to starch to complement or strengthen the advantages of each modification method.

References

Atwell WAPRaDL, Minneapolis MN, Hood LF, Lineback DR, Varriano-Marston E, Zobel HF (1988) The terminology and methodology associated with basic starch phenomena. Cereal Food World 33:306–311

Balmayor ER, Tuzlakoglu K, Azevedo HS, Reis RL (2009) Preparation and characterization of starch-poly-ε-caprolactone microparticles incorporating bioactive agents for drug delivery and tissue engineering applications. Acta Biomater 5:1035–1045. https://doi.org/10.1016/j.actbio.2008.11.006

BeMiller JN (2011) Pasting, paste, and gel properties of starch–hydrocolloid combinations. Carbohydr Polym 86:386–423

Benedetti G, Rastelli F, Damiani S, Calandri C, Crinò L (2004) Challenging problems in malignancy: case 1. Presentation of small-cell lung cancer with marked hyperamylasemia. J Clin Oncol 22:3826–3828

Bie P, Chen L, Li X, Li L (2016) Characterization of concanavalin A-conjugated resistant starch acetate bioadhesive film for oral colon-targeting microcapsule delivery system. Ind Crops Prod 84:320–329. https://doi.org/10.1016/j.indcrop.2016.02.023

Biliaderis CG (2009) Structural transitions and related physical properties of starch (Chap. 8). In: Whistler JBaR (ed) Starch (Chemistry and technology). 3rd edn. Academic Press, San Diego, pp 293–372. https://doi.org/10.1016/B978-0-12-746275-2.00008-2

Buléon A, Colonna P, Planchot V, Ball S (1998) Starch granules: structure and biosynthesis. Int J Biol Macromol 23:85–112

Cai L, Shi Y-C (2010) Structure and digestibility of crystalline short-chain amylose from debranched waxy wheat, waxy maize, and waxy potato starches. Carbohydr Polym 79:1117–1123

Cai L, Shi Y-C, Rong L, Hsiao BS (2010) Debranching and crystallization of waxy maize starch in relation to enzyme digestibility. Carbohyd Polym 81:385–393

Casas M, Ferrero C, de Paz MV, Jiménez-Castellanos MR (2009) Synthesis and characterization of new copolymers of ethyl methacrylate grafted on tapioca starch as novel excipients for direct compression matrix tablets. Eur Polymer J 45:1765–1776

Castro JV, Ward RM, Gilbert RG, Fitzgerald MA (2005) Measurement of the molecular weight distribution of debranched starch. Biomacromolecules 6:2260–2270

Chan HT, Leh CP, Bhat R, Senan C, Williams PA, Karim AA (2011) Molecular structure, rheological and thermal characteristics of ozone-oxidized starch. Food Chem 126:1019–1024

Constantin M, Fundueanu G, Cortesi R, Esposito E, Nastruzzi C (2003) Aminated polysaccharide microspheres as DNA delivery systems. Drug Deliv 10:139–149. https://doi.org/10.1080/10717540390215537

Cummings J, Englyst H (1991) Measurement of starch fermentation in the human large intestine. Can J Physiol Pharmacol 69:121–129

da Rosa Zavareze E, Pinto VZ, Klein B, El Halal SLM, Elias MC, Prentice-Hernández C, Dias ARG (2012) Development of oxidised and heat–moisture treated potato starch film. Food Chem 132:344–350

Dandekar P, Jain R, Stauner T, Loretz B, Koch M, Wenz G, Lehr C-M (2012) A hydrophobic starch polymer for nanoparticle-mediated delivery of docetaxel. Macromol Biosci 12:184–194. https://doi.org/10.1002/mabi.201100244

Desai KGH (2005) Preparation and characteristics of high-amylose corn starch/pectin blend microparticles: a technical note. AAPS PharmSciTech 6:E202–E208. https://doi.org/10.1208/pt060230

Duan S-H, Xu D-Q, Miao M (2007) Review on advance of modified starch based on physical methods. Food Sci 28:361–366

Elvira C, Mano JF, San Roman J, Reis RL (2002) Starch-based biodegradable hydrogels with potential biomedical applications as drug delivery systems. Biomaterials 23:1955–1966. https://doi.org/10.1016/s0142-9612(01)00322-2

Englyst HN, Kingman S, Cummings J (1992) Classification and measurement of nutritionally important starch fractions. Eur J Clin Nutr 46:S33–S50

Fichter M, Baier G, Dedters M, Pretsch L, Pietrzak-Nguyen A, Landfester K, Gehring S (2013) Nanocapsules generated out of a polymeric dexamethasone shell suppress the inflammatory response of liver macrophages. Nanomed Nanotechnol Biol Med 9:1223–1234. https://doi.org/10.1016/j.nano.2013.05.005

Fonseca LM, Gonçalves JR, El Halal SLM, Pinto VZ, Dias ARG, Jacques AC, da Rosa Zavareze E (2015) Oxidation of potato starch with different sodium hypochlorite concentrations and its effect on biodegradable films. LWT-Food Sci Technol 60:714–720

Gallant DJ, Bouchet B, Baldwin PM (1997) Microscopy of starch: evidence of a new level of granule organization. Carbohyd Polym 32:177–191

Goszczynski TM, Filip-Psurska B, Kempinska K, Wietrzyk J, Boratynski J (2014) Hydroxyethyl starch as an effective methotrexate carrier in anticancer therapy. Pharmacol Res Perspect 2:e00047–e00047. https://doi.org/10.1002/prp2.47

Haralampu SG (2000) Resistant starch—a review of the physical properties and biological impact of RS3. Carbohydr Polym 41:285–292. https://doi.org/10.1016/S0144-8617(99)00147-2

Henrist D, Lefebvre R, Remon JP (1999) Bioavailability of starch based hot stage extrusion formulations. Int J Pharm 187:185–191

Holm J, Björck I, Ostrowska S, Eliasson A-C, Asp N-G, Larsson K, Lundquist I (1983) Digestibility of Amylose-Lipid Complexes in-vitro and in-vivo Starch-Stärke. 35:294–297. https://doi.org/10.1002/star.19830350902

Huang Y, Liu M, Gao C, Yang J, Zhang X, Zhang X, Liu Z (2013) Ultra-small and innocuous cationic starch nanospheres: preparation, characterization and drug delivery study. Int J Biol Macromol 58:231–239. https://doi.org/10.1016/j.ijbiomac.2013.04.006

Jiang G, Qiu W, DeLuca PP (2003) Preparation and in vitro/in vivo evaluation of insulin-loaded poly (acryloyl-hydroxyethyl starch)-PLGA composite microspheres. Pharm Res 20:452–459. https://doi.org/10.1023/a:1022668507748

Kim H-Y, Park SS, Lim S-T (2015) Preparation, characterization and utilization of starch nanoparticles. Colloids Surf B Biointerfaces 126:607–620. https://doi.org/10.1016/j.colsurfb.2014.11.011

Kittipongpatana OS, Burapadaja S, Kittipongpatana N (2009) Carboxymethyl Mungbean starch as a new pharmaceutical gelling agent for topical preparation. Drug Dev Ind Pharm 35:34–42. https://doi.org/10.1080/03639040802144229

Kuakpetoon D, Wang YJ (2001) Characterization of different starches oxidized by hypochlorite. Starch–Stärke 53:211–218

Kumar P, Ganure AL, Subudhi BB, Shukla S (2015) Preparation and characterization of pH-sensitive methyl methacrylate-g-starch/hydroxypropylated starch hydrogels: in vitro and in vivo study on release of esomeprazole magnesium. Drug Deliv Transl Res 5:243–256

Larionov NV, Ponchel G, Duchene D, Larionova NI (1999) Biodegradable cross-linked starch/protein microcapsules containing proteinase inhibitor for oral protein administration. Int J Pharm 189:171–178. https://doi.org/10.1016/s0378-5173(99)00249-5

Lefnaoui S, Moulai-Mostefa N (2011) Formulation and in vitro evaluation of kappa-carrageenan-pregelatinized starch-based mucoadhesive gels containing miconazole. Starch-Stärke 63:512–521. https://doi.org/10.1002/star.201000141

Lefnaoui S, Moulai-Mostefa N (2014) Investigation and optimization of formulation factors of a hydrogel network based on kappa carrageenan–pregelatinized starch blend using an experimental design. Colloids Surf A 458:117–125. https://doi.org/10.1016/j.colsurfa.2014.01.007

Li Y, Zhang Z, Van Leeuwen HP, Cohen Stuart MA, Norde W, Kleijn JM (2011) Uptake and release kinetics of lysozyme in and from an oxidized starch polymer microgel. Soft Matter 7:10377–10385. https://doi.org/10.1039/c1sm06072d

Li G et al (2014) Hydroxyethyl starch conjugates for improving the stability, pharmacokinetic behavior and antitumor activity of 10-hydroxy camptothecin. Int J Pharm 471:234–244. https://doi.org/10.1016/j.ijpharm.2014.05.038

Li K et al (2015a) Mulberry-like dual-drug complicated nanocarriers assembled with apogossypolone amphiphilic starch micelles and doxorubicin hyaluronic acid nanoparticles for tumor combination and targeted therapy. Biomaterials 39:131–144. https://doi.org/10.1016/j.biomaterials.2014.10.073

Li M, Witt T, Xie F, Warren FJ, Halley PJ, Gilbert RG (2015b) Biodegradation of starch films: the roles of molecular and crystalline structure. Carbohyd Polym 122:115–122

Li G, Cai C, Qi Y, Tang X (2016) Hydroxyethyl starch–10-hydroxy camptothecin conjugate: synthesis, pharmacokinetics, cytotoxicity and pharmacodynamics research. Drug Deliv 23:277–284

Lu D, Xiao C, Xu S (2009) Starch-based completely biodegradable polymer materials Express. Polym Lett 3:366–375

Mahkam M (2010) Modified chitosan cross-linked starch polymers for oral insulin delivery. J Bioact Compatible Polym 25:406–418. https://doi.org/10.1177/0883911510369038

Marsh RDL, Blanshard JMV (1988) The application of polymer crystal growth theory to the kinetics of formation of the B-amylose polymorph in a 50% wheat-starch gel. Carbohydr Polym 9:301–317. https://doi.org/10.1016/0144-8617(88)90048-3

Michailova V, Titeva S, Kotsilkova R, Krusteva E, Minkov E (2001) Influence of hydrogel structure on the processes of water penetration and drug release from mixed hydroxypropylmethyl cellulose/thermally pregelatinized waxy maize starch hydrophilic matrices. Int J Pharm 222:7–17. https://doi.org/10.1016/S0378-5173(01)00706-2

Mua J, Jackson D (1997) Fine structure of corn amylose and amylopectin fractions with various molecular weights. J Agric Food Chem 45:3840–3847

Namazi H, Belali S (2016) Starch-*g*-lactic acid/montmorillonite nanocomposite: synthesis, characterization and controlled drug release study. Starch-Stärke 68:177–187. https://doi.org/10.1002/star.201400226

Oechslein CR, Fricker G, Kissel T (1996) Nasal delivery of octreotide: absorption enhancement by particulate carrier systems. Int J Pharm 139:25–32. https://doi.org/10.1016/0378-5173(96)04569-3

Oladebeye AO, Oshodi AA, Amoo IA, Karim AA (2013) Functional, thermal and molecular behaviours of ozone-oxidised cocoyam and yam starches. Food Chem 141:1416–1423

Pereswetoff-Morath L (1998) Microspheres as nasal drug delivery systems. Adv Drug Deliv Rev 29:185–194. https://doi.org/10.1016/S0169-409X(97)00069-0

Pérez S, Bertoft E (2010) The molecular structures of starch components and their contribution to the architecture of starch granules: a comprehensive review. Starch–Stärke 62:389–420

Qiao L, Gu Q-M, Cheng H (2006) Enzyme-catalyzed synthesis of hydrophobically modified starch. Carbohyd Polym 66:135–140

Ramasubbu N, Paloth V, Luo Y, Brayer GD, Levine MJ (1996) Structure of human salivary α-amylase at 1.6 A resolution: Implications for its role in the oral cavity. Acta Crystallogr Sect D: Struct Biol 52:435–446. https://doi.org/10.1107/S0907444995014119

Saboktakin MR, Maharramov A, Ramazanov MA (2009) pH-sensitive starch hydrogels via free radical graft copolymerization, synthesis and properties. Carbohydr Polym 77:634–638. https://doi.org/10.1016/j.carbpol.2009.02.004

Sadeghi M, Hosseinzadeh H (2008) Synthesis of starch-poly(sodium acrylate-co-acrylamide) super-absorbent hydrogel with salt and pH-responsiveness properties as a drug delivery system. J Bioact Compatible Polym 23:381–404. https://doi.org/10.1177/0883911508093504

Saikia C, Das MK, Ramteke A, Maji TK (2016) Effect of crosslinker on drug delivery properties of curcumin loaded starch coated iron oxide nanoparticles. Int J Biol Macromol 93:1121–1132. https://doi.org/10.1016/j.ijbiomac.2016.09.043

Sánchez-Rivera M, García-Suárez F, Del Valle MV, Gutierrez-Meraz F, Bello-Pérez L (2005) Partial characterization of banana starches oxidized by different levels of sodium hypochlorite. Carbohyd Polym 62:50–56

Santander-Ortega M et al (2010) Nanoparticles made from novel starch derivatives for transdermal drug delivery. J Control Release 141:85–92

Sanz T, Salvador A, Baixauli R, Fiszman S (2009) Evaluation of four types of resistant starch in muffins. II. Effects in texture, colour and consumer response. Eur Food Res Technol 229:197–204

Shaikh MMM, Gramopadhye NS, Wardole AA, Lonikar SV (2015) Starch-acrylic acid hydrogel: synthesis, characterization and drug release study. World J Pharm Pharm Sci 4:942–954

Sieradzki R, Traitel T, Goldbart R, Geresh S, Kost J (2014) Tailoring quaternized starch as a non-viral carrier for gene delivery applications. Polym Adv Technol 25:552–561. https://doi.org/10.1002/pat.3277

Simi CK, Emilia Abraham T (2007) Hydrophobic grafted and cross-linked starch nanoparticles for drug delivery. Bioprocess Biosyst Eng 30:173–180. https://doi.org/10.1007/s00449-007-0112-5

Singh AV, Nath LK, Guha M (2011) Synthesis and characterization of highly substituted acetylated moth bean starch. J Polym Mater 28:277–285

Situ W, Chen L, Wang X, Li X (2014) Resistant starch film-coated microparticles for an oral colon-specific polypeptide delivery system and its release behaviors. J Agric Food Chem 62:3599–3609. https://doi.org/10.1021/jf500472b

Situ W, Li X, Liu J, Chen L (2015) Preparation and characterization of glycoprotein-resistant starch complex as a coating material for oral bioadhesive microparticles for colon-targeted polypeptide delivery. J Agric Food Chem 63:4138–4147

Sivapragasam N, Thavarajah P, Ohm JB, Thavarajah D (2014) Surface properties of semi-synthetic enteric coating films: Opportunities to develop bio-based enteric coating films for colon-targeted delivery. Bioact Carbohydr Dietary Fibre 4:139–143. https://doi.org/10.1016/j.bcdf.2014.09.001

Smith BW, Roe JH (1949) A photometric method for the determination of α-amylase in blood and urine, with use of the starch-iodine color. J Biol Chem 179:53–59

Soares GA, Castro ADD, Cury BSF, Evangelista RC (2013) Blends of cross-linked high amylose starch/pectin loaded with diclofenac. Carbohyd Polym 91:135–142. https://doi.org/10.1016/j.carbpol.2012.08.014

Sudo K, Kanno T (1976) Properties of the amylase produced in carcinoma of the lung. Clin Chim Acta 73:1–12

Surini S, Anggriani V, Anwar E (2009) Study of mucoadhesive microspheres based on pregelatinized cassava starch succinate as a new carrier for drug delivery. J Med Sci 9:249–256. https://doi.org/10.3923/jms.2009.249.256

Tako M (1996) Molecular origin for the thermal stability of waxy–rice (Kogane) starch. Starch-Stärke 48:414–417

Tan Y, Xu K, Li Y, Sun S, Wang P (2010) A robust route to fabricate starch esters vesicles. Chem Commun 46:4523–4525. https://doi.org/10.1039/C000471E

Tatongjai J, Lumdubwong N (2010) Physicochemical properties and textile utilization of low- and moderate-substituted carboxymethyl rice starches with various amylose content. Carbohydr Polym 81:377–384
Tester RF, Karkalas J, Qi X (2004) Starch—composition, fine structure and architecture. J Cereal Sci 39:151–165
Thiele C, Loretz B, Lehr C-M (2017) Biodegradable starch derivatives with tunable charge density—synthesis, characterization, and transfection efficiency. Drug Deliv Transl Res 7:252–258. https://doi.org/10.1007/s13346-016-0333-8
Thompson DB (2000) Strategies for the manufacture of resistant starch. Trends Food Sci Technol 11:245–253
Tuma RF, Forsberg JO, Agerup B (1982) Enhanced uptake of actinomycin D in the dog kidney by simultaneous injection of degradable starch microspheres into the renal artery. Cancer 50:1–5
Tuovinen L, Peltonen S, Liikola M, Hotakainen M, Lahtela-Kakkonen M, Poso A, Jarvinen K (2004a) Drug release from starch-acetate microparticles and films with and without incorporated alpha-amylase. Biomaterials 25:4355–4362. https://doi.org/10.1016/j.biomaterials.2003.11.026
Tuovinen L et al (2004b) Starch acetate microparticles for drug delivery into retinal pigment epithelium - in vitro study. J Control Release 98:407–413. https://doi.org/10.1016/j.jconrel.2004.05.016
Vanier NL, da RosaZavareze E, Pinto VZ, Klein B, Botelho FT, Dias ARG, Elias MC (2012) Physicochemical, crystallinity, pasting and morphological properties of bean starch oxidised by different concentrations of sodium hypochlorite. Food Chem 131:1255–1262
Wang S, Copeland L (2013) Molecular disassembly of starch granules during gelatinization and its effect on starch digestibility: a review. Food Func 4:1564–1580
Wang Q, Hu X, Du Y, Kennedy JF (2010) Alginate/starch blend fibers and their properties for drug controlled release. Carbohydr Polym 82:842–847. https://doi.org/10.1016/j.carbpol.2010.06.004
Wang S, Blazek J, Gilbert E, Copeland L (2012) New insights on the mechanism of acid degradation of pea starch. Carbohyd Polym 87:1941–1949. https://doi.org/10.1016/j.carbpol.2011.09.093
Wepner B, Berghofer E, Miesenberger E, Tiefenbacher K, Ng PNK (1999) Citrate starch—application as resistant starch in different food systems. Starch-Stärke 51:354–361. https://doi.org/10.1002/(sici)1521-379x(199910)51:10%3c354:aid-star354%3e3.0.co;2-w
Wurzburg O (1986) Converted starches. Modified Starches: Prop Uses 18–41
Wuttisela K, Triampo W, Triampo D (2009) Chemical force mapping of phosphate and carbon on acid-modified tapioca starch surface. Int J Biol Macromol 44:86–91
Xiao C, Fang F (2009) Ionic self-assembly and characterization of a polysaccharide-based polyelectrolyte complex of maleic starch half-ester acid with chitosan. J Appl Polym Sci 112:2255–2260. https://doi.org/10.1002/app.29763
Xiao H, Yang T, Lin Q, Liu GQ, Zhang L, Yu F, Chen Y (2016) Acetylated starch nanocrystals: preparation and antitumor drug delivery study. Int J Biol Macromol 89:456–464. https://doi.org/10.1016/j.ijbiomac.2016.04.037
Xu E-Y, Guo J, Xu Y, Li H-Y, Seville PC (2014) Influence of excipients on spray-dried powders for inhalation. Powder Technol 256:217–223. https://doi.org/10.1016/j.powtec.2014.02.033
Yamada H, Loretz B, Lehr C-M (2014) Design of starch-graft-PEI polymers: an effective and biodegradable gene delivery platform. Biomacromolecules 15:1753–1761
Yoon H-S, Lee J, Lim S-T (2009) Utilization of retrograded waxy maize starch gels as tablet matrix for controlled release of theophylline. Carbohydr Polym 76:449–453
Zhang SD, Zhang YR, Zhu J, Wang XL, Yang KK, Wang YZ (2007) Modified corn starches with improved comprehensive properties for preparing thermoplastics. Starch–Stärke 59:258–268
Zhang AP et al (2013a) Disulfide crosslinked PEGylated starch micelles as efficient intracellular drug delivery platforms. Soft Matter 9:2224–2233. https://doi.org/10.1039/c2sm27189c
Zhang Z, Shan HL, Chen L, He CL, Zhuang XL, Chen XS (2013b) Synthesis of pH-responsive starch nanoparticles grafted poly (L-glutamic acid) for insulin controlled release. Eur Polym J 49:2082–2091. https://doi.org/10.1016/j.eurpolymj.2013.04.032

Chapter 4
Starch-Based DDSs with Stimulus Responsiveness

Abstract Owing to the excellent biodegradability and biocompatibility of starch, numerous efforts have been undertaken to prepare stimulus-responsive drug delivery systems (DDSs) from starch and its derivatives during the past few decades. The biological stimuli at the different organ and cellular compartment-specific levels or pathological conditions including pH, enzyme, temperature, and redox potential have been exploited for the development of starch-based stimulus-responsive DDSs. These types of stimulus responsiveness of starch-based DDSs can be achieved by incorporating functional groups to starch, such as disulfide bonds to acquire redox sensitiveness, or by changing physicochemical properties, such as hydrophilicity/hydrophobicity of starch derivatives to acquire temperature sensitiveness. Besides, magnetic-responsive starch-based DDSs have been developed by the incorporation of magnetic particles with starch film coatings or in starch matrix. The individual starch-based, stimulus-responsive DDSs have to some extent been reasonably well validated. Furthermore, two or more response elements have been combined to functionalize starch-based DDSs for smart drug release behavior. In this chapter, the role and application of starch and its derivatives in DDSs endowed with individual, dual, and multi-stimuli responsiveness will be discussed.

Keywords Starch-based drug delivery system · Stimulus responsiveness · pH responsiveness · Enzymatic responsiveness · Temperature responsiveness · Redox responsiveness · Magnetic responsiveness · Dual and multi-responsiveness

4.1 Introduction

Medicines rely on the use of drugs to manage and treat diseases, especially cancers, which are one of the principal causes of morbidity and mortality worldwide (Allen and Cullis 2004). However, drugs, having a beneficial action, can also exhibit side effects that may limit their clinical application. For example, cytotoxic compounds used in cancer therapy can not only kill target cells but also normal cells in the body, resulting in undesired side effects. In the ideal case, drugs would be applied in vivo at exactly the therapeutic concentration and would precisely target cells that cause

J. Chen et al., *Drug Delivery Applications of Starch Biopolymer Derivatives*
https://doi.org/10.1007/978-981-13-3657-7_4

diseases. However, the drug release is difficult to control. The insufficient therapeutic treatment and the side effects (e.g., toxicity, anti-inflammatory response) associated with pharmaceutics remain to be challenged.

There has long been a desire to achieve the delivery of therapeutics to various sites in the body to maximize therapeutic potential and minimize any side effects. This ambition would rely on the improved understanding of the physiological and pathological bases, as well as the advances in polymer materials development. Recently, DDSs have been designed by using a wide array of strategies for achieving better therapeutic application. One of the extraordinary examples is stimulus-responsive DDS. Controlled release of therapeutics can be triggered by various biological or external stimuli such as temperature, pH, and hypoxia at the organ, tissue, and cell levels (Kommareddy and Amiji 2005; Shenoy et al. 2005; Shenoy and Amiji 2005), or by the effects of electric/magnetic fields or light (Jie et al. 2013; Mamada et al. 1990; Shiga et al. 1992). The use of stimulus-responsive carriers offers an interesting opportunity where the delivery system becomes an active participant, rather than a passive vehicle, in the optimization of therapy. The benefits of stimulus-responsive DDSs are especially important when the stimuli are unique to the disease pathology, allowing the carriers to respond specifically to the pathological "triggers".

Materials in the application of drug delivery need to minimize the possibilities of hypersensitivity reactions and afford excellent tissue compatibility (Shi et al. 2010). Owing to the motivating properties of starch such as biocompatibility and nontoxicity, the interest in the development of starch as smart materials has strongly grown in recent years (Kim et al. 2015). Besides, unlike charged polysaccharides, such as pectin (Thakur et al. 1997), alginate (Haug and Larsen 1963) and gelatin (Cortesi et al. 1998), the solubility of starch shows no pH-dependent behavior, which favors the stability of starch-based carriers. In addition, due to the presence of various derivatizable groups on molecular chains, starch can be easily modified, resulting in different enzymatic resistance and various starch-based functional materials. Based on those advantages, starch has a long history of the application in DDSs as excipients. Some examples include Starch 1500® (partially hydrolyzed starch) as fillers (Bolourchian et al. 2009; Das and Das 2004), sodium starch glycolate as a disintegrant (Faroongsarng and Sukonrat 2008; Galgatte et al. 2013), and cold water-soluble starch (also sometimes referred to as drum-dried, roll-dried, cold water swellable or pregelatinized starch) as binders (Callens and Remon 2000; Codd and Deasy 1998).

With the advance of DDSs and modification technologies, starch has been investigated not only as simply excipients but also as, with the adjusted molecular structure, polymers to control drug release. Besides, the development in the field of medicine has seen the ongoing transformation of starch-based DDSs to be therapies based on nucleic acids, immunological antibodies, proteins, and peptides (Fig. 4.1). These bioactive substances, which can be much larger than small drug molecules (e.g., doxorubicin (DOX), an anticancer drug), and significantly more sensitive to environmental effects, will require adequate protection and specificity. This is undoubtedly a challenging issue since the required starch-based DDSs is expected to be on-demand control of therapeutics. Different starch-based DDSs have been developed, and these DDSs can be classified according to the different carrier strategies (Fig. 4.1). The

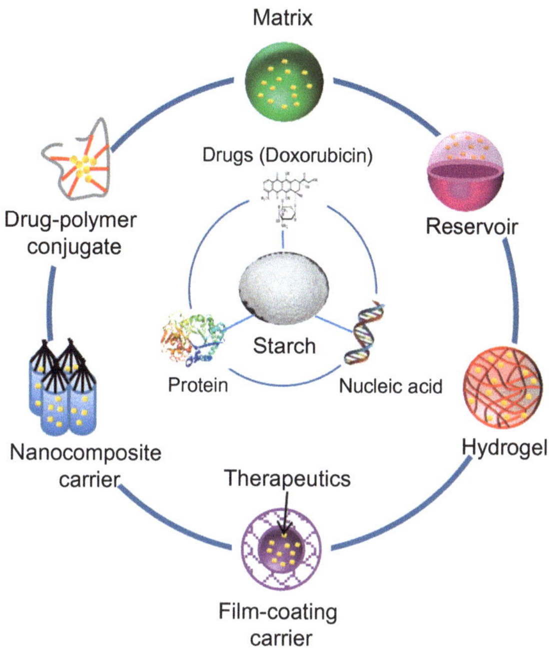

Fig. 4.1 Therapeutics that starch-based drug delivery systems are applied to and different strategies to construct starch-based drug carriers

controlled release of drugs can be achieved by using diffusion-controlled matrices and reservoirs, enzymatically or chemically regulated biodegradable and bioerodible materials, and solvent-activated hydrogels. In the case of starch-based DDSs, the controlled release operates through a combination of two or more these mechanisms. In a matrix-based DDS, the drug diffuses through a tortuous network of interconnected pores; however, in a reservoir, the drug is encapsulated and passes through a semipermeable membrane. In a film-coating DDS, the drug is released throughout a semipermeable membrane or when the film degrades. Hydrogel-based DDS releases the drug through a constrained network whose mesh size depends on hydration and polymer architecture. Recently, new starch-based carriers have emerged with better controlled-release performance, such as nanocomposite carrier (Fig. 4.1). For example, as a typical nanocomposite carrier, gated mesoporous silica material controls the drug release when the gated material responds to the environment or degrades (Acosta et al. 2014). In addition, drug–polymer conjugates can be formed through chemical bonds or physical drug–polymer interactions, and these conjugates can release the drug through the breakage of those interactions.

To achieve controlled and sustained drug release, starch-based DDSs have been most widely investigated in traditional dosage forms [e.g., tablets (van Veen et al. 2005), capsules (Misic et al. 2014), films (Karrout et al. 2011), pellets (Freire et al. 2010), and beads (Setty et al. 2014)]. In recent years, more functional and complicated dosage forms of starch emerged and became a trend, such as hydrogels (Pourjavadi et al. 2013), MPs (Grinberg and Gedanken 2010), NPs (Shalviri 2012), nanocom-

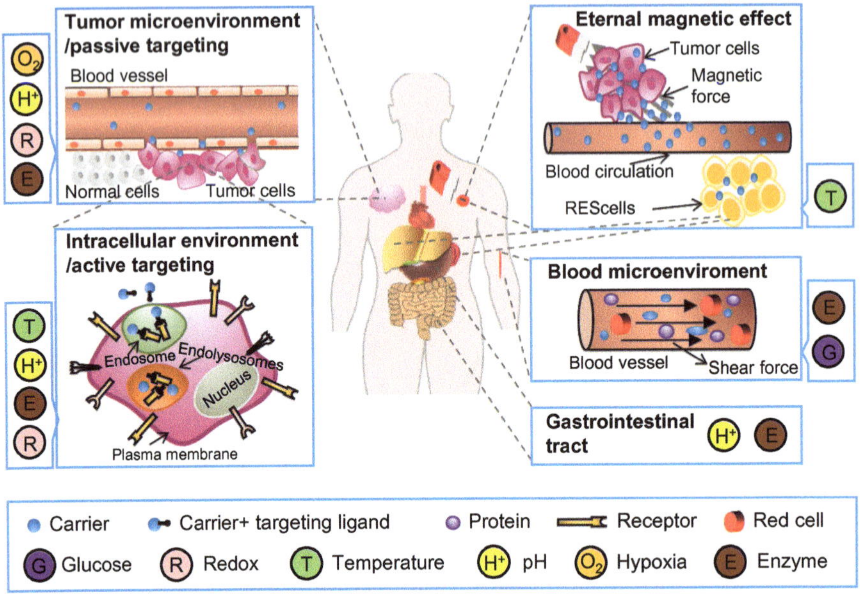

Fig. 4.2 Typical physiological environments with associated biological stimuli and interactions applied in starch-based drug delivery system. For target-specific delivery to tumor sites, three ways can be adapted: one uses internal physiological environments (passive and active targeting); the other uses external stimuli (e.g., magnetic force targeting to avoid the clearance of reticuloendothelial system (RES) cells). After drug-loaded nanoparticles are injected into the blood circulation system, they are internalized through endocytosis and remained in the early endosome, which then mature and differentiate into late endosomes. Late endosomes can fuse with lysosomes, forming endolysosomes

posites (Subramanian et al. 2014), and micelles (Yang et al. 2014c). This book aims at summarizing the developments in starch polymers used in advanced drug dosage forms (hydrogel and micro- and nanoparticulate carriers). Up to date, there have been some reviews concerning the application of starch in DDSs to form extended-release tablets (Hong et al. 2016), nanocarriers (Kim et al. 2015; Le Corre et al. 2010; Rodrigues and Emeje 2012), hydrogels (Ismail et al. 2013; Zhang et al. 2005), and its application in oral site-specific drug delivery (Watts and Smith 2005) and wound dressing (Torres et al. 2013). To the best of our knowledge, no advanced starch-based DDSs that possess smart or active drug release behavior have been reviewed. With improved understanding of the physiological and pathological bases, the role of starch and its derivatives in the stimulus-responsive starch-based DDSs based on the biological stimuli at the different organ and cellular compartment-specific levels and physical stimuli are particularly focused (Fig. 4.2). The guiding principles for effective advanced starch-based DDSs and how these principles are informing for the development of starch-based DDSs are highlighted and illustrated as follows.

4.2 pH

The physiological pH triggers in starch-based DDSs can be classified into two main categories: (a) the pH shifts at the organ levels (especially the GIT) or pathological conditions, and (b) the cellular compartment-specific pH changes (Table 4.1).

4.2.1 pH Shifts in the GIT

Starch-based carriers (mainly hydrogels) that are pH-responsive at the organ levels are mostly for oral administration, especially oral site-specific DDSs (Abdel Ghaffar et al. 2016; Bajpai and Saxena 2004a, b; Desai 2007; Kumar et al. 2015; Liu et al. 2011; Rodrigues and Emeje 2012; Saboktakin et al. 2011; Tan et al. 2016; Zhang et al. 2009). These oral DDSs are of high potential for the delivery of nutraceuticals, drugs, and proteins to the lower part of the GIT (Table 4.1). To achieve this goal, carriers have to pass through the stomach, the upper part of the intestine and must use the pH characteristics of the intestine and the colon to specifically release therapeutics in the lower part of the GIT. The easiest way to meet the pH or ionic strength changes of the external environment along the GIT is to bring ionic groups (usually carboxyl groups) to starch through one or two modified methods (Table 4.1). The behavior of starch-based, pH-sensitive DDSs will be discussed as follows according to their pH-triggered changes of carriers. Also, factors that affect drug release will be addressed since the behavior of drug release is complex and multifactorial and can be regulated by a variety of physicochemical properties.

4.2.1.1 Swelling/Shrinkage

With a chemically cross-linked network structure, starch-based hydrogels, in most cases, are only capable of swelling, not dissociation or degradation, despite its high hydrophilicity and the electrostatic repulsions between ionized starch-based polymers responding to the environment. In this case, the swelling/shrinkage change of carriers responding to the pH changes along the GIT will control the drug release to achieve site-specific delivery (Chap. 2, Fig. 2.1).

Among all modifications, grafting copolymerization of vinyl monomers onto starch is the most common approach to bring anionic groups and a firm cross-linked network to starch hydrogels or hydrogel-based MPs or NPs (Table 4.1). The pH-dependent feature of polyvinyl polymers can endow starch with pH responsiveness along the GIT. For instance, polyacrylic acid (PAA) hydrogels swell significantly above pH 5 while do not swell significantly below pH 4, since the pKa of acrylic acid (AA) is between 4.5 and 5.0, which is accord with the pH changes of the GIT (Baker et al. 1994). Polyacrylonitrile (PAN) (Dragan and Apopei 2013; Sadeghi and Hosseinzadeh 2008), poly(methyl methacrylate) (PMMA) (Kumar et al. 2015), poly-

Table 4.1 Examples of starch-based carriers responding to the pH changes of organ levels, pathological conditions, and cellular compartments

Dosage form	Therapeutics	Modified starch	Release behavior	References
Oral delivery along gastrointestinal tract (GIT)				
Swelling shrinkage				
		Chemical cross-linking:		
Hydrogels	–	High amylose/pectin mixtures cross-linked with sodium trimetaphosphate (STMP) and dried by oven or lyophilization	Lyophilization samples presented higher swelling ability than oven samples and showed more swelling capacity in high pH medium	Soares et al. (2013a)
Semi-interpenetrating polymer networks hydrogels	5-fluorouracil	Cross-linked poly(aspartic acid) (PAsp) hydrogel, starch as entrapping polymer	The amount of drug (92%) released in a pH 7.4 medium was higher than that in a pH 2.2 (only about 20%) medium within 7 h Drug release rate increased with the increased PAsp content	Liu et al. (2011)
		Grafting copolymerization-free radical polymerization:		
Hydrogels	Sodium diclofenac	Acrylamide and acrylic acid grafted with starch/ethylene-*co*-vinyl alcohol copolymer blends	The maximum hydration degree was at pH = 7.4, and the lowest was observed in an acidic pH medium	Elvira et al. (2002)
Hydrogels	5-aminosalicylic acid	Carboxymethyl starch (CMS, degree of substitution (DS): 0.49) grafted with poly(methacrylic acid) (PMAA)	The drug rate increased at pH 7.4, but decreased at pH 1 The drug release increased with the increased incorporation of PMAA and the reduced particle size	Saboktakin et al. (2007, 2009a)

(continued)

Table 4.1 (continued)

Dosage form	Therapeutics	Modified starch	Release behavior	References
Hydrogels	Insulin	Grafting copolymerization of PMAA and poly(ethylene glycol) (PEG) monomethyl ether methacrylate onto chitosan cross-linked CMS	A higher release rate at pH 7.4 than at pH 1.2 As the PMAA content increased, the degradation of insulin decreased in both acidic and basic fluids	Mahkam (2010)
Hydrogels	Amoxicillin trihydrate	Starch-*g*-poly(sodium-2-acrylamido-2-methyl-1-propane sulfonate-*co*-methacrylic acid)	The amount of drug release in pH 9.2 buffer solution was found to be higher than that in pH 7.0 and pH 4.0 buffer solutions	Clara and Natchimuthu (2017)
Microspheres	Curcumin	Grafting glycidyl methacrylate modified starch microspheres prepared by water/oil emulsion	After 12 h, in simulated gastric fluid (SGF), only 18% of encapsulated drugs was released, increasing up to 95% by 96 h; in simulated intestinal fluid (SIF), about 14% was released, increasing up to 38% by 48 h, and then a drop in the concentration up to 20% was observed for 96 h	Pereira et al. (2013)
		Grafting copolymerization-direct gamma radiation:		
Hydrogels	Vitamin B12	*N*-vinyl pyrrolidone and starch grafted with acrylic acid	The minimum release was at a acid medium (pH 1.1), while high release rate was at pH 6.8	Eid (2008)
		Grafting copolymerization-in situ cross-linking:		

(continued)

Table 4.1 (continued)

Dosage form	Therapeutics	Modified starch	Release behavior	References
Hydrogels	Ibuprofen	Grafting polyacrylonitrile (PAN) chains by the starch to obtain starch–poly(sodium acrylate-*co*-acrylamide)	At low pH values, the gel swelling decreased and minimized drug release; in alkaline media, the gel swelling degree increased and thus drug release increased	Sadeghi and Hosseinzadeh (2008)
Dissociation, dissolution, and **degradation**				
		Physical cross-linking:		
Hydrogels	Salicylic acid	Alginate and starch fiber prepared by coagulation with $CaCl_2$	The release rate was maximal at pH 7.4 and minimum at pH 1.0 The released amount of drug was increased with an increased proportion of starch, decreased as the amount of drug loaded increased, but the cumulative release amount increased	Wang et al. (2010)
		Blending:		
Microcapsules	Oyster peptide	Alginate/chitosan/high amylose corn starch (HACS) powder prepared using external or internal emulsion gelation method	Compared with external gelation microcapsules, the drug release of internal gelation microcapsules was faster	Zhang et al. (2009)
Nanocomposite films	Curcumin	Starch/chitosan/montmorillonite blended nanocomposite films	More than 80% curcumin was released from composites at pH 7.4 within 8 h, whereas less than 44% of drugs was released at pH 3.4 within 4 h	Mohanty et al. (2015)

(continued)

Table 4.1 (continued)

Dosage form	Therapeutics	Modified starch	Release behavior	References
Microparticles	Diclofenac sodium	HACS/pectin blend microparticles prepared by spray-drying conditions	More pectin ratio, less cumulative amount of drug release; more HACS ratio, more drug release due to the porous structure	Desai (2005, 2007)
Micelles	–	**Grafting**: starch cellulose acetate *co*-acrylate	The amount released depended on the pH of the media in the following order: basic media > acidic media > neutral Sustained release of drug extended to about 20 days	Helaly et al. (2013)
Drug–polymer interaction				
Microgels	Lobular protein lysozyme	2,2,6,6-tetramethyl-1-piperidinyloxy (TEMPO)-mediated oxidation of starch and cross-linked by STMP	Swelling increased with the increased pH and remained constant from pH 5 to 8 Increasing the salt concentration or the pH caused more mobile fractions of protein; thus, more drug released	Li et al. (2009, 2011a, b)
Pathological conditions and cellular compartment				
Degradation				
Nanocrystals	Doxorubicin (DOX)	Using acetic anhydride as acetylating agent through reaction with starch nanocrystals prepared by sulfuric acid hydrolysis, DS: 0.04, 0.08, and 0.14	Significantly faster drug release was found at pH 5.7 compared with pH 6.5 and 7.4	Xiao et al. (2016)

(continued)

Table 4.1 (continued)

Dosage form	Therapeutics	Modified starch	Release behavior	References
Nanocomposite films	Curcumin	Cassava starch nanoparticles prepared by acid hydrolysis and then blending with poly(vinyl alcohol)	More drug release was found at an acid medium; sustained release was found at pH 7.4 Faster release rates were found with the increased amount of drug loading	Athira and Jyothi (2015)
Degradation-pH-sensitive bond				
Nanoparticles	DOX	Layer-by-layer assembly of chitosan/dialdehyde starch polyelectrolyte multilayers onto mesoporous silica nanoparticles surface	After 72 h, about 34.25% of the DOX was released in the pH 5.0, while less than 9% of the DOX was released in the pH 7.4	Wang et al. (2014)
Micelles	DOX	The amphiphilic polymer prepared by grafting hydrophobic deoxycholic acid into mPEGylated starch	At pH 5.5, the drug release rate was faster than that at pH 7.4; at pH 7.4, diffusion-controlled drug release, while at pH 5.5, both diffusion and macromolecular relaxation were major factors for drug release	Yang et al. (2014c)

(continued)

Table 4.1 (continued)

Dosage form	Therapeutics	Modified starch	Release behavior	References
Drug–polymer interaction				
Nanoparticles	DOX	PMAA-polysorbate 80-grafted starch	At the same drug loading (i.e., 11% DOX), DOX release from the nanoparticles was much faster at pH 5 than that at pH 7.4 This nanoparticle overcame multidrug resistance in human breast cancer cells, thus significantly improved the cell killing efficacy of DOX	Shalviri et al. (2013, Shalviri (2012)
Degradation and drug–polymer interaction				
Graphene nanosheets	Hydroxycamptothecin hydrochloride	Starch functionalized graphene nanosheets via the reduction of exfoliated graphene oxides by soluble starch	At pH 7.4, the release rate was very slow and only about 3% of the total amount was released; with the decrease of pH value, the release rate became fast gradually	Liu et al. (2015)

methacrylic acid (PMAA) (Clara and Natchimuthu 2017; Güler et al. 2015; Mahkam 2010; Saboktakin et al. 2007, 2009a), 2-hydroxyethyl methacrylate (HEMA) (Gils et al. 2009) and PAA (Abdel Ghaffar et al. 2016; Bajpai and Saxena 2004a, b; Eid 2008; Elvira et al. 2002; Gils et al. 2009; Shaikh et al. 2015) have been frequently grafted onto starch to obtain carboxylate or carboxamide groups. Free radical polymerization is a commonly used grafting method to prepare pH-sensitive starch polymers (Table 4.1). Also, other alternative modification methods, such as direct gamma radiation technique (Abdel Ghaffar et al. 2016), have been developed, due to their advantages such as easier process control and environment friendliness (El-Arnaouty et al. 2015a; b). Interestingly, a new synthesis method for pH-sensitive hydrogels was discovered by using PAN to be grafted onto starch. The nitrile groups of PAN were converted to a mixture of hydrophilic carboxamide and carboxylate groups during alkaline hydrolysis followed by *in situ* cross-linking of the grafted PAN chains (Sadeghi and Hosseinzadeh 2008). This simple physical mixture of starch and PAN yielded starch–poly(sodium acrylate-*co*-acrylamide) superabsorbent hydrogels. These hydrogels showed a pH-responsive swelling–deswelling behavior at pH 2 and 8, and the release rate was much faster at pH 7.4 than at pH 1.2 (Sadeghi and Hosseinzadeh 2008).

The swelling behavior of carriers for oral DDSs is highly dependent on the pH shifts of the GIT, and the drug release is in accordance with the swelling capacity, which means that the more swelling of the carrier, the higher amount of the drug is released (Abdel Ghaffar et al. 2016; Bajpai and Saxena 2004b; Clara and Natchimuthu 2017; Dragan and Apopei 2013; Eid 2008; Elvira et al. 2002; Güler et al. 2015; Gils et al. 2009; Kumar et al. 2015; Liu et al. 2011; Mahkam 2010; Mohanty et al. 2015; Saboktakin et al. 2007, 2009a; Sadeghi and Hosseinzadeh 2008; Shaikh et al. 2015; Soares et al. 2013a, b; Tan et al. 2016). At low pH, the H^+ ions presented in the medium diffuse into the polymer matrix through water-filled microspores in the fully hydrated gel matrix. These H^+ ions protonate the $-COO^-$ groups to yield unchanged $-COOH$ groups. This change ultimately results in the folding of the macromolecular chains due to the formation of hydrogen bonding between $-COOH$ and $-OH$ groups in the hydrogel network. The folding of chains minimizes the release of drug molecules from the polymer matrix. However, when the hydrogel is placed in a medium with high pH, the almost complete ionization of $-COOH$ groups presented within the polymer network not only increases the ion osmotic swelling pressure to a great extent but also enhances the relaxation of macromolecular chains because of the repulsion among similarly charged $-COO^-$ groups. These two factors ultimately result in a greater increase in the water uptake and thus enhance drug release. The sharp swelling/shrinkage on-off switching behavior of the hydrogels makes them suitable candidates for oral site-specific DDSs. For example, the grafting copolymerization of methyl methacrylate onto starch and hydroxypropyl starch (HPS) made these two starch derivatives suitable for intestinal drug delivery. In a pH 6.8 solution, the drug release was found to be sustained up to 14 h with 80–90% drug release, while only 40–45% of the drug release was observed in a pH 1.2 medium (Kumar et al. 2015). Well correlated to that, the in vitro and in vivo analysis showed that the amount of radioactive tracer released from the labeled hydrogel was minimal when the hydrogel was in the stomach, whereas it increased rapidly as the hydrogel reached the small intestine (Kumar et al. 2015).

4.2.1.2 Dissociation, Dissolution, and Degradation

For those carriers where chemical cross-linking is absent, the intermolecular interactions are not robust enough to counteract the electrostatic repulsions to keep the carrier intact. Thus, the dissociation of starch-based carriers occurs due to the highly unstable hydrogen bonding and strong electrostatic repulsion of ionic polymers in different pH media. The micelle of starch polymers is a typical example that can experience the dissociation of its conformation due to its pH responsiveness (Chap. 2, Fig. 2.1). More specifically, the pH affects not only the degree of ionization but also the hydrophilicity of micelles, which both contribute to the self-assembly/disassembly behavior of micelles. For example, under an acidic condition (pH 5.0), poly(L-glutamic acid) blocks of the micelle prepared by grafting starch with poly(L-glutamic acid) as grafting chains were nearly unchanged, resulting in the enhanced hydrophobic association between the hydrogen bonding of poly(L-glutamic acid) blocks (–COOH) and starch blocks (–OH) (Zhang et al. 2013c). Meanwhile, the self-assembled aggregates became compact, resulting in a smaller size of the micelles, compared to the size at pH 6.8. While in a basic medium, the –COOH groups of poly(L-glutamic acid) became $-COO^-$ groups, increasing the dissolution of the aggregates and allowing the drug to diffuse out easily (Zhang et al. 2013c). Thus, all the samples exhibited a lower cumulative drug release percentage in the simulated gastric fluid (SGF) (pH 1.2) than in the simulated intestinal fluid (SIF) (pH 6.8) (Zhang et al. 2013c).

In some cases, degradation of starch-based carriers also occurs at the same time, which can result from the susceptibility and instability of starch polymers in a certain range of pH media. Dissolution often comes along dissociation and/or degradation because of the different solubility of ionized starch derivatives themselves or their degraded polymers in different pH media (Chap. 2, Fig. 2.1). Those carriers can be built simply by blending raw or modified starches with ionic polymers or grafted starch polymers (Table 4.1). Among these blended polymers, a typical example is polyelectrolyte complex (PEC), which is made of two oppositely charged substances (Saboktakin et al. 2011; Zhang et al. 2009). The pH-responsive drug release relies on the ionization of its carboxylic groups which results in high solubility of the polymer matrix in the basic media (Lima-Tenório et al. 2015). The PEC NPs prepared by a complex coacervation process of chitosan–carboxymethyl starch (CMS) showed high potential as colon-specific DDS, with $85.2 \pm 4.0\%$ of drugs released after 5 h (Saboktakin et al. 2011). Moreover, another noteworthy example is blending high amylose cornstarch (HACS) with ionized polymers to achieve more controlled drug release, due to the unique feature of HACS. HACS has a high content of amylose, which is insoluble in both acidic and neutral environments (Noakes et al. 1996). An effort has been made to prepare noncross-linked HACS controlled-release formulation by blending gelatinized HACS with pectin (Desai 2005, 2007). The advantage of blending HACS with pectin is that it can decrease the dissolution of pectin in the gastric environment, thus, compared with pectin MPs, decreasing the release rate of the drug in an SGF (Desai 2005). Besides, the introduction of HACS to HACS/pectin blended MPs decreased the drug dissolution in the gastric condition thus decreased the drug release (Desai 2005). These tableted gelatinized HACS/pectin MPs showed more drug release in a neutral solution (Desai 2007) since pectin aggregates tend to dissociate and dissolve, attributed to the ionization of its carboxylic groups and the higher solubility of the pectin at this pH (Liu et al. 2003).

Improved controlled drug release behavior of starch-based, pH-responsive DDSs can be achieved by sophisticated and combined form of starch carriers. For instance, the incorporation of the drug-loaded chitosan microcapsules into the starch film resulted in a synergistic effect in pH responsiveness and achieved long-term drug release (Fig. 4.3b). In acidic (pH 3.5 and 5.5) and alkaline media (pH 8.5 and 10.5), low or high pH weakened the hydrogen bonding between starch polymers, which was the sole structural support of the starch-based blended film. Moreover, the weakened hydrogen bonding accelerated the degradation of the film, causing a massive drug release (Fig. 4.3a). However, things were different for chitosan microcapsules. Under acidic conditions, the protonation of amino groups could increase the swelling of chitosan microcapsules, which enhanced the rate of drug release. While in an alkaline medium, as the swelling of microcapsules could be suppressed by the low solubility of the chitosan, the massive drug release was reduced to an extent comparable to that in the acidic media. In a neutral medium (pH 7.4), the hydrogen bonding of the film matrix was relatively stable, and the amino groups of the chitosan could not be protonated, which allowed the drugs to be released through the compound matrix moderately. Thus, the rate of drug release in different pH media can be ranked in the order, pH 3.5 > pH 10.5 $\gg$ pH 5.5 > pH 8.5 > pH 7.4 (Fig. 4.3c) (Huo et al. 2016).

4.2.1.3 Drug–Polymer Interactions and Others

The interactions between the drug and starch polymers in the starch-based matrix can also be a way to achieve pH responsiveness. This response behavior can be a result of the changes of hydrogen bonding responding to the pH shifts. For example, when curcumin was loaded into starch NPs prepared by ultrasonication, the drug release of curcumin in the SIF was faster than that in the SGF (Li et al. 2016). The different rate of drug release can be attributed to the hydrogen bonding between the –OH groups of curcumin and the –OH groups of glucose units of starch as influenced by the environmental pH (Jin et al. 2006). In the SIF, the –OH group would more likely to dissociate, which led to the breakage of the hydrogen bonding. Meanwhile, the charge repulsion of starch polymers was increased, and all of this conduced to rather faster drug release in the SIF (Li et al. 2016).

In particular, for protein drugs, drug–starch interactions to achieve pH responsiveness are much more widespread. The charge of the drug and the carrier, as well as the resulting electrostatic interactions between them, can respond to pH (Chap. 2, Fig. 2.1). Besides, these drug–starch interactions help the carrier to improve its drug loading capacity. The latter is particularly important in developing DDSs since in

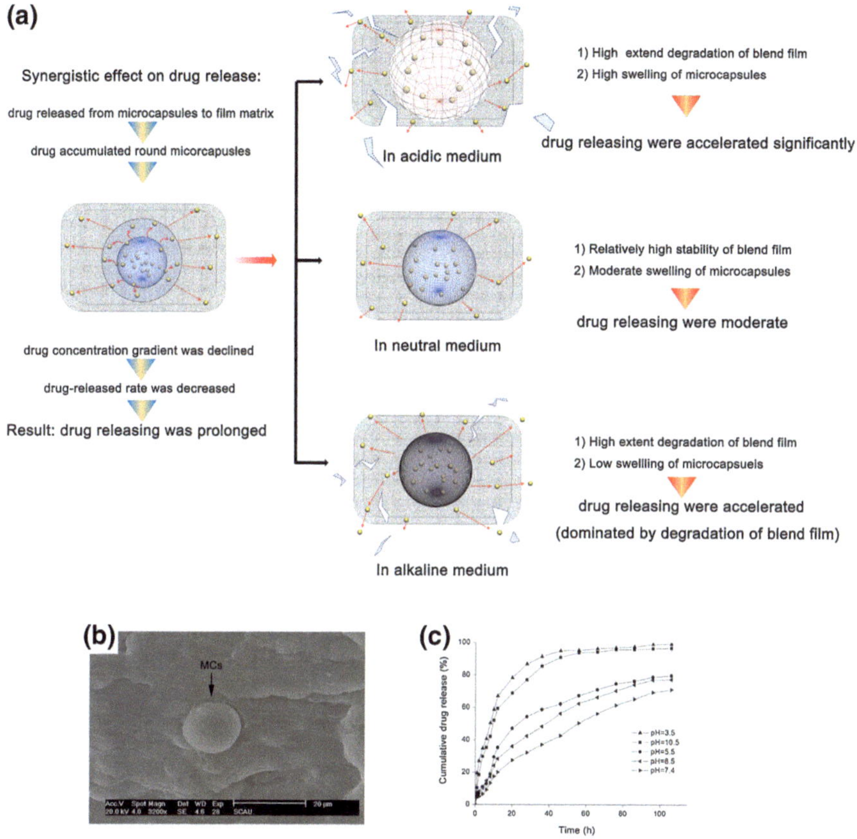

Fig. 4.3 Drug release behavior of chitosan microcapsules (CS-MCs)/starch-based blended film. **a** Scheme of drug release mechanism of drug-loaded CS-MCs and starch films; **b** SEM images of cross section of CS-MCs and starch films; and **c** release profile of drug-loaded CS-MCs/starch-based blended film in media with different pH values. Reprinted from Huo et al. (2016), Copyright (2016), with permission from Elsevier

most cases, a high drug loading capacity is difficult to achieve. The protein drugs are loaded to starch-based carriers by electrostatic interactions, and the charge of the protein determine the binding affinity of the protein in the carriers. For instance, the lysozyme loading content in the oxidized starch microgels was found to increase with an increasing degree of oxidation of starch in a certain range of concentration (0.1–0.2 mg/mL) (Li et al. 2009). In the pH range of 3–9, the lysozyme (the isoelectric point is ~10) was positive and the gel was negative (Li et al. 2011b), and high degree oxidation resulted in high charge densities of starch, thus high electrostatic attractions and a high drug loading content. High salt concentration and high pH led to weakened electrostatic interactions, resulting in more mobile fractions of lysozyme thus more protein drug release. It can be implied that the proteins embedded in the gel can be protected when the salt concentration and pH are kept low, while triggered to be released at high salt concentrations and high pH values (Li et al. 2011a).

Besides the mechanisms discussed in Chap. 2, Fig. 2.1, more interestingly, other mechanisms have also been found to achieve pH responsiveness of starch-based carriers. It was reported that simply using trisodium citrate as the cross-linker to prepare a starch-based hydrogel can also achieve pH responsiveness (Abhari et al. 2017). The rate of drug release into the salt-free water was pH-dependent and occurred faster at neutral pH (pH 7.0) than under acidic conditions (pH 2.0) (Abhari et al. 2017). For its pH sensitivity, the author explained that hydronium and chlorine ions exist at pH 2.0 progressively lost their translational entropy due to the diffusion of drugs from the hydrogel into the surrounding water. Hence, the drug release from the hydrogel was slower into the acidic absorbent than the neutral counterpart likely due to entropic effects. This shows that there are still some blank pages to draw to achieve pH sensitivity of starch-based carriers.

4.2.1.4 Factors That Affect Drug Release

Up to date, most starch-based pH-sensitive carriers were reported to be successful for the delivery of drugs to the lower GIT, especially to the colon (Abdel Ghaffar et al. 2016; Bajpai and Saxena 2004b; Desai 2005, 2007; Liu et al. 2011; Saboktakin et al. 2011; Tan et al. 2016). Only a few showed high amounts of drug release under acidic conditions, due to the high solubility of the drugs (Pereira et al. 2013) or the degradation of starch (Athira and Jyothi 2015) in an acidic medium. These previous studies suggested that the behavior of drug release is complex and multifactorial, and can be regulated by a variety of physicochemical properties. The drug release behavior can be manipulated by polymer and carrier properties (e.g., nature, composition, the cross-linking degree, the pore size of carriers), the chemical structure of the drug, drug–polymer affinity, and environment condition (Table 4.1).

Concerning the properties of starch polymers, the nature of starch polymers has a great impact on the drug release behavior. Normally, higher hydrophobicity of starch polymers favors slower drug release behavior of carriers. For example, the HPS-*g*-PMMA hydrogel, due to the higher hydrophobicity of HPS, showed a slower drug release rate than starch-*g*-PMMA (Kumar et al. 2015). In the case of interpenetrating polymer networks (IPNs) hydrogels, the swelling and drug release behavior of the hydrogel can be affected by the nature of the entrapped starch polymer significantly. When using potato starch-*g*-PAN as an entrapped polymer of semi-IPNs hydrogels, compared with potato starch, before alkaline hydrolysis, the swelling was more affected by pH and potato starch-*g*-PAN had lower swelling capacity than potato starch due to its anionic carboxylate groups. However, after hydrolysis, it was higher than potato starch (Dragan and Apopei 2013). This change was contributed by the decreased concentration of –OH groups in the interior of the gel with the presence of $–COO^-$ groups in the case of the anionic polyelectrolyte (starch-*g*-PAN) as the entrapped polymer (Dragan et al. 2012; Zhao et al. 2010, 2011).

The percentage of composition of starch/polymers in the blends also has a huge impact on the material properties and drug release behavior. An increase in the starch content in the polymer blends usually leads to an increase in the drug release due to the hydrophilicity of starch (Gils et al. 2009). However, there are mixed results and optimal composition ratios of starch polymers have to be found out for better performance of the DDSs. The tableted MPs based on 1:3 or 3:1 (HACS:pectin) blend ratios were found to be better formulations since they exhibited effective sustained drug release, compared to other formulations (Desai 2007). Beyond a certain range of concentration of starch/polymers, the decreased swelling behavior occurred, thus the drug release decreased. This can be related to the interactions of starch itself, such as the retrogradation of HACS, which can make the structure more rigid and tortuous (Zhang et al. 2009), or the interactions between starch and polymers in the blends. The increased cross-linking and hydrogen bonding between the starch and carboxyl polymers caused by the high content of anionic polymers leads to the decreased swelling (Abdel Ghaffar et al. 2016; Clara and Natchimuthu 2017; Eid 2008; Gils et al. 2009; Mahkam 2010; Saboktakin et al. 2009a; Sadeghi and Hosseinzadeh 2008). Thus, an appropriate compositions ratio will be highly critical for the tight control of hydrogel swelling and drug release.

For cross-linked hydrogels, the cross-linking degree has an enormous impact on the swelling properties of the hydrogel. In most cases, a higher cross-linking degree hinders the diffusion of the solvent into the network, hence lowering the swelling ratio (Table 4.1). Thus, increasing the amount of the cross-linker for the starch modification could decrease the drug release. Besides, the cross-linking reaction can result in lower average pore size within hydrogels and a greater extent of compaction of the matrix, which may be another reason for the low swelling. The average pore size is a limiting factor of drug release because of its direct relation to the capacity of hydrogels to retain water (Wu et al. 2008). In the case of semi-IPNs hydrogels of starch/poly(aspartic acid) compounds, the swelling capacity and the drug release content were increased with a higher average pore diameter (Liu et al. 2011). This average pore diameter was observed to be increased with the increased poly(aspartic acid) content, owing to the more hydrophilic groups (–COOH and –CONH–) in poly(aspartic acid), which accordingly increased the swelling ratio (Liu et al. 2011).

For drug-loaded carriers, the cumulative release amount and the drug release rate could be increased with the increased percentage of drug loading (Athira and Jyothi 2015; Mohanty et al. 2015), due to the high concentration gradient of drugs between the matrix and the environment. However, other researchers (Wang et al. 2010) reported that the increased amount of drug loading could lead to a decreased drug release rate. This suggests the drug release behavior of carriers is highly dependent on the properties of drugs, and the solubility of the drug was found to affect the drug release behavior significantly (Pereira et al. 2013; Surini et al. 2009; Tan et al. 2016). One example is that the high solubility of propranolol hydrochloride at pH 1.2 gives the pregelatinized cassava starch succinate microspheres a pH-sensitive release behavior, leading to a faster drug release of the microspheres at pH 1.2 than at pH 7.2 (Surini et al. 2009). Another example is that, despite that, the swelling degree of hydrogels was minimum under acidic conditions, there was a sharp increase in the drug release because of the high solubility of ciprofloxacin hydrochloride at this pH (Hui et al. 2012). In contrast, at pH 7.4, the hydrogel was only able to release about 50% of the drug because of the low aqueous solubility of ciprofloxacin hydrochloride, suggesting that this DDS was suitable for gastric drug delivery (Hui et al. 2012).

Concerning about the environmental factors, the swelling of carriers in the presence of various salt solutions at different concentrations is important when they are used for the application in the biomedical field. For example, physically cross-linked hydrogels (pea starch/calcium alginate hydrogels formed by hydrogen bonding and ionic interactions) was found to be sensitive to the salt concentrations due to the osmotic pressure-induced gel swelling (water in and ions out) (Mehyar et al. 2008). Usually, the pH-responsive carriers are also sensitive to the ions in the swelling medium, and the presence of ions has a profound effect on the absorbency behavior (Abhari et al. 2017; Clara and Natchimuthu 2017; Dragan and Apopei 2013; Gils et al. 2009; Li et al. 2009, 2011a, b; Sadeghi and Hosseinzadeh 2008; Wang et al. 2010). Increasing the ionic mobile concentration difference between the polymer gel and external medium will decrease the gel swelling capacity due to the drop of the ionic pressure. Results showed that independent of the radius of the cation or the nature of the anion, all univalent cationic species of the same concentration have a similar effect on the water absorbency (Gils et al. 2009). Also, in the case of salt solutions with multivalent cations, "ionic cross-linking" at the surface of particles causes an appreciable decrease in swelling capacity. With an increased cationic charge, the cross-linking is increased, which consequently decreases the swelling (Sadeghi and Hosseinzadeh 2008).

4.2.2 Pathological Conditions and Cellular Compartment-Specific pH Changes

A faster drug release from starch polymeric systems at moderately acidic pH is desirable for the effective delivery of anticancer drugs to the solid tumor. One way to achieve this goal is to take advantage of the enhanced acidic environment of tumor tissues and cellular endosomes/lysosomes. The carriers are expected to release more drugs in tumor tissue than in the systemic circulation. As a result, lower systemic toxicity and higher therapeutic efficacy could be obtained.

The pH responsiveness for cancer treatment can also result from the conformation changes of micro- or nanoparticulate of starch-based carriers, such as swelling, dissolution, degradation, and drug–polymer interaction at lower pH (Chap. 2, Fig. 2.1). Classic examples of using solubility or degradation in response to the local pH level are those carriers made from starch nanocrystals prepared by acid hydrolysis, such as acetylated starch nanocrystals, owing to the inclination of starch to degrade under acidic conditions (Xiao et al. 2016). The large surface of the smaller sized nanocrystals and the highly hydrophilic hydroxyl groups of starch enhanced biodegradation rates and promoted water absorption and the drug diffusion, which accordingly enhanced the drug release (Xiao et al. 2016).

Alternatively, DDSs with acid-labile bonds, which can be broken under mildly acidic conditions, can deliver drugs for cancer treatment. The Schiff base bonds (imine groups) have been applied to be cross-linking bonds in starch organic NPs or inorganic nanomaterials for the pH-triggered release of therapeutic drugs (Table 4.1). The imine group was labile under acid conditions, and the stability of these bonds decreased as the pH decreased (Wang et al. 2014). For example, the versatile hydrogel fabricated by the *in situ* cross-linking of starch NPs (decorated with cholesterol groups and aldehyde groups) with polyvinylamine has been exploited to be tumoral drug delivery carrier materials (Li et al. 2014). These carriers had a very slow drug release rate at pH 7.4 (only 3% of the DOX released within 36 h), but a very fast drug release rate in an acid medium (85.9% of DOX released at pH 5.0, and 90.3% at pH 3.0) (Li et al. 2014). These results could be explained by the hydrolysis under acid conditions of the bonds formed by the Schiff base reaction between the amino group of the DOX and the aldehyde of the modified starch.

Besides the Schiff base bonds, ester bonds have also been used for the drug release triggered by the endosomal acid pH (Table 4.1). A novel amphiphilic polymer was successfully prepared by grafting hydrophobic deoxycholic acid onto starch, which can self-assemble into micelles with spherical core–shell structures (Yang et al. 2014a). The pH sensitivity of starch–deoxycholic acid conjugates was attributed to the introduced ester linkages between deoxycholic acid and starch (Fig. 4.4a). Because of the hydrolytic susceptibility of ester linkage and the electrostatic repulsion between deoxycholic acids [pKa = 6.58 (Chen et al. 2013)] under acidic environments, the particle size and the CMC of this micelle were decreased with higher pH (Fig. 4.4b). Moreover, under acidic conditions, the reduced hydrogen bonding between the hydroxyl groups of starch and deoxycholic acid, which decreased the hydrophobic ability, also contributed to a high CMC value (Yang et al. 2014c). All these factors led to the loose self-assembled structure of micelles under acidic conditions, which resulted in more drug release.

By modulating the properties of starch polymers and choosing appropriate drugs, the electrostatic interactions between the drug and starch can also be used to achieve pH responsiveness in an acidic environment. One typical example is using cationic starch NPs for the delivery of different proteins (e.g., bovine serum albumin (BSA), and bone morphogenetic protein-4 (BMP-4)) (Huang et al. 2013). Various drug–starch polymer interactions resulted in different drug loading levels and release behaviors in solutions (pH 6.0) (Fig. 4.5a). The pKa of BSA is 4.8, and can be negatively charged at pH 6.0 due to acid dissociation; on the contrary, BMP-4 would be positively charged at pH 6.0 when its pKa is > 6.0 (Huang et al. 2013). As a result, BSA would be absorbed onto the cationic NPs surface and into their interior via electrostatic attraction, and thus a high loading content was achieved. In contrast, BMP-4 showed a lower drug loading ratio due to the electrostatic repulsion (Fig. 4.5b). However, if the pH of the medium was lower than the pKa of BSA, which could be achieved in the lysosome of cells, the drug release will be faster owing to the electrostatic repulsion since BSA became negatively charged (Huang

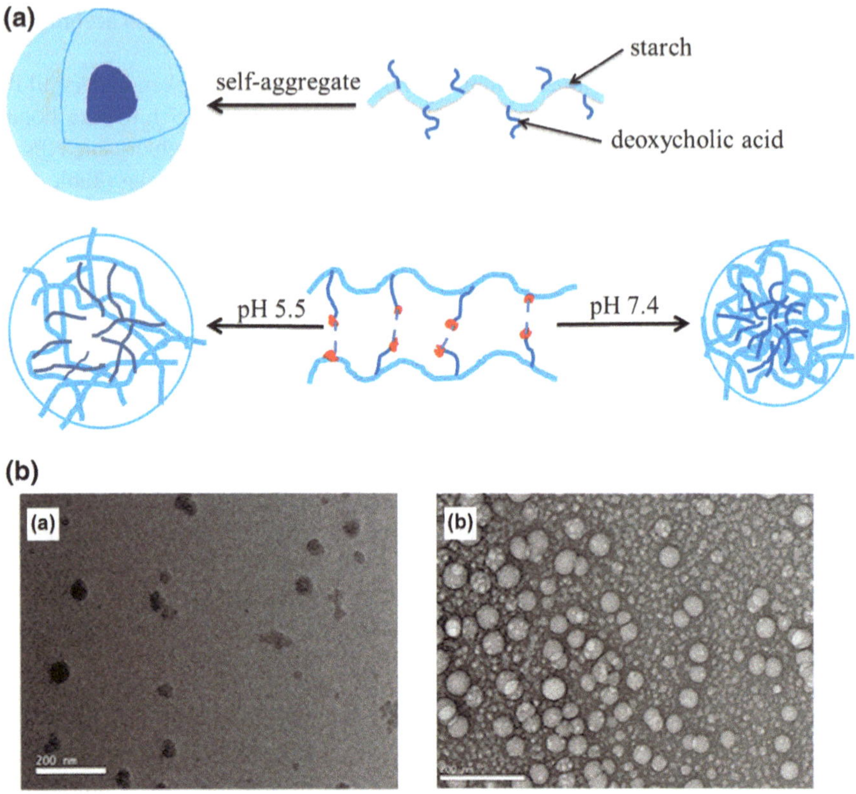

Fig. 4.4 Self-aggregation behaviors of starch–deoxycholic acid (St–DCA) conjugates. **a** Schematic illustration of self-aggregation behaviors of St–DCA conjugates under different pH conditions. **b** TEM images of self-aggregation nanoparticles of St–DCA conjugates in different pH solutions: (a) St–$DCA_{8.9}$ (pH 7.4) and (b) St–$DCA_{8.9}$ (pH 5.5) (8.9 refers to the degree of substitution of DCA per 100 glucose units of starch). Reprinted from Yang et al. (2014a), Copyright (2014), with permission from Elsevier

et al. 2013). These results suggest that the cationic starch particles possess good capacity in delivering the negatively charged molecules to cancer cells.

Like oral pH-responsive starch-based DDSs, for their responsiveness to the changes in tumor and lysosomal pH, the drug release of starch-based DDSs can be modulated by several factors such as polymer composites and drug properties. However, unlike the consistency of drug release with the swelling capacity in the case of oral DDSs, the more swelling capacity of starch-based DDSs did not result in more drug release. On the contrary, more drug release occurred with carriers with less swelling capacity under acidic conditions (Li et al. 2014; Shalviri et al. 2013; Shalviri 2012). Those phenomena can be explained by taking other factors into account, which outweigh the effect of swelling capacity. For instance, the stability of drug–polymer interactions (whether physically or chemically), such as the

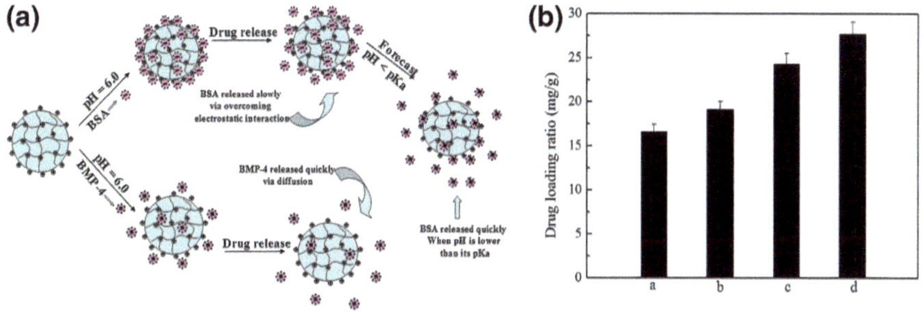

Fig. 4.5 Drug release behaviors of cationic starch nanospheres. **a** The diagrammatic process of drug loading and release; **b** Drug loading ratio of bovine serum albumin (BSA) and bone morphogenetic protein-4 (BMP-4) (a, $Particle_{10.2}$–BMP-4; b, $Particle_{10.2}$–BSA; c, $Particle_{6.8}$–BSA; and d, $Particle_{3.4}$–BSA) (3.4, 6.8, and 10.2 refer to the mole ratios of $POCl_3$ based on the mole amount of starch structure units). Reprinted from Huang et al. (2013), Copyright (2013), with permission from Elsevier

DOX–PMAA interactions (Shalviri et al. 2013; Shalviri 2012), or the Schiff base bonds between the amino group of the DOX and the aldehyde of starch NPs (Li et al. 2014), was liable under acidic conditions. This could result in more drug release in an acid medium, even though lower swelling capacity compared with in a basic environment.

4.3 Enzymes

Biodegradability within the body is generally required for all DDS carriers. The enzymatic degradation of starch can be either used as a trigger to release drugs or be modified to inhibit its digestion to extend the drug release at the site of action. As mentioned in Sect. 3.3, α-amylase found in the pancreatic juices (Ramasubbu et al. 1996) and blood (Smith and Roe 1949), or microbial enzymes secreted by the colon microorganisms (Yang et al. 2002), can also be a good target for starch-based degradable DDSs. Also, the existence of amylase-producing tumors has been reported (Hata et al. 1990; Joseph et al. 1992; Zhang et al. 2013a). Physiological enzyme triggers for starch-based DDSs can be classified into three broad categories: (a) enzymes in the GIT; (b) blood serum endoamylase; and (c) cellular compartment enzymes (Table 4.2).

There are mixed results of enzymatic degradation behaviors of starch-based carriers. It is well established that, with the increased content of enzyme in the external solution, the degradation could be enhanced (Bajpai and Saxena 2006; Bajpai et al. 2006; Hamdi et al. 1998). The enzymatic degradation of starch polymers was reported to be almost linear over time (Fig. 4.6a, b), indicating a "zero-order" release of reducing sugars (Bajpai and Saxena 2006; Bajpai et al. 2006; Huang et al. 2013). Besides, in the case of cross-linked starch microspheres, except for 5 IU/l of α-amylase con-

Table 4.2 Examples of starch-based carriers degraded by enzymes of organ levels, blood serum endoamylase, and cellular compartment enzymes

Dosage form	Therapeutics	Modified starch	Behavior of enzymatic degradation and drug release	References
Enzymes in the gastrointestinal tract (GIT)				
Pancreatic amylase in the intestine				
Microgels	Lysozyme	Oxidation mediated starch then cross-linked by sodium trimetaphosphate	Protein release rate: slow in the first 300s, increased after about 500s, and completed and reached a plateau after complete amylase degradation	Li et al. (2011b)
Microspheres	–	Cross-linked starch using epichlorohydrin	Total degradation time within 150 min; degradation time decreased with the increasing enzymatic activity	Hamdi et al. (1998)
Microcapsules	Aprotinin-protease inhibitors	Starch/bovine serum albumin using interfacial cross-linking with terephthaloyl chloride	Microcapsule was mainly digested by α-amylase (0.2–1 mg/ml), not digested by esterase (19 IU/ml); a mixture of these two enzymes with pancreatin (1 mg/ml) only slightly enhanced the degradation rate	Larionov et al. (1999)

(continued)

Table 4.2 (continued)

Dosage form	Therapeutics	Modified starch	Behavior of enzymatic degradation and drug release	References
Ternary semi-interpenetrating network (IPNs) hydrogels	Vitamin B12	Grafting copolymerization of monomer methacrylamide, in the presence of starch and polyethylene glycol (PEG)	Degradation rate was almost linear over a time period of nearly 6 h, increased with the enzyme activity The degradation was enhanced with the increased content of starch and water content, and was lowered with an increased amount of PEG	Bajpai and Saxena (2006), Bajpai et al. (2006)
Amylose inclusion complexes	Ibuprofen	Enzymatic polymerization of amylose in the presence of ibuprofen	V-type crystalline Lower in simulated gastric fluid (SGF) and simulated intestine fluid (SIF), the drug release was activated by enzymatic digestion	Carbinatto et al. (2016), Yang et al. (2013)

(continued)

Table 4.2 (continued)

Dosage form	Therapeutics	Modified starch	Behavior of enzymatic degradation and drug release	References
	Nimesulide (NMS) and praziquantel (PZQ)	Preparation of NMS-amylose and PZQ-amylose in the presence of palmitic acid (PA) Drug–polymer ratio (1:5, 1:15, or 1:30) and temperature of incubation (60°C or 90 °C)	High yield of amylose inclusion complexes (> 57%) and drug content (up to 68.16%) The highest enzymatic resistance of PZQ1:30PA60°C complex made the release time longer and the full release of PZQ occurred at 240 min, whereas the complexes with NMS and PZQ1:5PA90°C did it in 60 min	Carbinatto et al. (2016)

(continued)

Table 4.2 (continued)

Dosage form	Therapeutics	Modified starch	Behavior of enzymatic degradation and drug release	References
Microbial enzymes in the colon				
Semi-IPN hydrogels	Riboflavin	Free radical grafting copolymerization of starch-*g*-poly(acrylic acid)	More amount (nearly 75%) of drug released was observed in enzyme-containing media than the blank (nearly 39%) Degradation rate was almost linear with time for at least 6 h and enhanced with the increased content of starch and initial water; An increase in cross-linking degree reduced enzymatic hydrolysis	Bajpai and Saxena (2004a, b)
Microspheres	Rofecoxib	Cross-linked starch using epichlorohydrin	Less susceptibility to microbial enzymes compared to native starch The release of rofecoxib from microspheres was found to be 40% in less than 50 min compared to the raw drug (99%)	Thombre et al. (2009)
Microcapsules	Fish oil	Casein-glucose-resistant starch (RS) blends (high amylose corn starch (HACS), Hylon VII)	Only 4–6% of oil was released in SGF and SIF	Patten et al. (2009)

(continued)

Table 4.2 (continued)

Dosage form	Therapeutics	Modified starch	Behavior of enzymatic degradation and drug release	References
Tableted microparticles	Diclofenac sodium	HACS (Hylon VII)/pectin blends	HACS and pectin at 1:3 and 3:1 ratios led to slower and effective controlled drug release	Desai (2005, 2007)
Film-coated microparticles	Insulin	RS was modified by a high-temperature–pressure reaction, enzymatic debranching, and retrogradation	RS content: 10.4–41.8% Release percentage of insulin was up to 25.31% within the first 8 h in the upper GIT and then reached 80.66% at 30 h	Situ et al. (2014)
Microspheres	5-Fluorouracil; metronidazole	Assam Bora rice starch	Insignificant release was found in the stomach and small intestine, and major drug release was found in the cercal content, while raw drug distributed predominantly in the upper GIT	Ahmad et al. (2012a, b, 2013)
Human serum endoamylase				
Microparticles	Dextran, lysozyme	Radical copolymerization of hydroxyethyl methacrylate–hydroxyethyl starch (HES); DS: 0.01–0.22	Hydrogel degradation accelerated the release Increasing DS reduced drug release and enzymatic degradation	Schwoerer et al. (2008, 2009)

(continued)

Table 4.2 (continued)

Dosage form	Therapeutics	Modified starch	Behavior of enzymatic degradation and drug release	References
Nanospheres	Nattokinase	Ultra-small and anionic starch nanospheres prepared by reverse micro-emulsion method using phosphoric trichloride as cross-linker	Improved stability of drugs; 23% of the total amount of drug injected was retained in the kidney one hour after injection of the drug-sphere combination	Huang et al. (2013)
Cellular compartment enzymes				
Nanopolycomplex	Plasmid DNA	Quaternization starch with 3-chloro-2-hydroxypropyltrimethylammonium chloride; nitrogen/phosphate atom (N/P) ratios: 1.5, 3, 5, 7	Higher N/P ratio and incubation time resulted in greater transfection efficiency Complexes of N/P = 5 entered 80% of the cells	Sieradzki et al. (2014)
	Plasmid DNA	Oxidized starch then conversed with alkyl diamines; DS (COOH): 5–62%, DS (amine): 5–61%	The speed of degradation was gradually decreased with increased spacer length and DS; for DS, dramatic change in degradation speed occurred at DS 20% to constant (DS: ~ 30%)	Thiele et al. (2017)
Microspheres	Plasmid DNA	Cross-linked starch followed by amination with *N,N*-diethyl-2-chloroethylamine or *N*-glycidyl-*N,N*-dimethyl-*N*-methylammonium chloride	An initial fast release of DNA (30 min) followed by slower release rate over 14 days More DNA was released with the increased NaCl concentration	Constantin et al. (2003)

(continued)

Table 4.2 (continued)

Dosage form	Therapeutics	Modified starch	Behavior of enzymatic degradation and drug release	References
Nanocomposites	Plasmid DNA	Starch-graft-polyethylenimine (PEI) polymers; N/P ratio: 10, 12, 14, 16	A linear degradation over the 90 min, followed by zero-order kinetics Degradation of starch-*g*-PEI was 1.7 − 2.2 times slower than that of raw starch Polymers with N/P = 14 showed the best transfection; the optimal of the best transfection efficiency: > 100 kDa starch and 30% of PEI contained α-amylase degradation greatly reduced cytotoxicity after the 3 h incubation	Yamada et al. (2014)
Nanodendrites	Small interfering RNA (siRNA)	Dendritic starch modified using glycidyltrimethylammonium chloride	Cationic dendritic starch-siRNA maintained sustained knockdown of the protein target for 6 days	Engelberth et al. (2015)

(continued)

Table 4.2 (continued)

Dosage form	Therapeutics	Modified starch	Behavior of enzymatic degradation and drug release	References
Nanoparticles	Rhodamine B	Esterification of waxy maize starch using palmitoyl chloride and acetic anhydride; DS (palmitoyl chloride): 0.005–0.30, DS (acetic anhydride): 2.51–2.78	Hydrolytic membrane degradation An immediate burst release phase was presented at the initial 3–4 h, and then sustained release followed	Tan et al. (2010)
Microparticles	Calcein	SA; DS = 2.6	The weight of SA in the film decreased 52–72% during 24 h incubation in retinal pigment epithelium D407 cells homogenate During 3 h incubation, 8.1 ± 0.8% of these cells for ocular treatment took up SA microparticles	Tuovinen et al. (2004b)

(continued)

Table 4.2 (continued)

Dosage form	Therapeutics	Modified starch	Behavior of enzymatic degradation and drug release	References
Starch–drug conjugates	10-hydroxy camptothecin (10-HCPT)	HES–10-HCPT conjugate; HES: *M*: 20 kDa; DS of HES = 0.5	Minimal (less than 20%) drug release was observed in 48 h at pH 6.0 and 7.4 medium but it was promoted by enzymes in plasma and liver homogenate A better antitumor efficiency against nude mouse with Hep-3B tumor than the commercial 10-HCPT injection Pharmacokinetics: half-life ($t_{1/2}$): 10-HCPT: about 10 min; 10-HCPT–HES conjugate: 4.38 h	Li et al. (2016)

centration, the linearized degradation profile was found out to be typically biphasic, a linear degradation followed by a plateau (Fig. 4.6c), of which the degradation rate was much lower (Hamdi et al. 1998). This suggests the nondegraded microspheres existed at that moment. However, the degradation of oxidized cross-linked microgels, which was detected by the filled protein release kinetics, was not linear: the release rate was slow initially, and after about 500s it increased, and then after 1200s, the drug release completed and reached a plateau when all the starch has been degraded (Fig. 4.6d) (Li et al. 2011b). All these results suggest that the enzymatic degradation of starch-based carriers is complicated, and may be particular to a specific starch-based carrier due to the different enzymatic resistance caused by inhomogeneous preparation conditions. Take biphasic degraded cross-linked starch microspheres for example, one assumption of the nonuniform degradation is that RS fractions exist in the microspheres due to the inhomogeneity caused by the cross-linking process (Hamdi et al. 1998). Another explanation relates to the size polydispersity of microsphere resulting from the preparation. More specifically, as the microspheres of small mean diameters were degraded totally, large microspheres left was degraded slowly due to the low surface-to-volume ratio (Hamdi et al. 1998).

This part of the chapter only focuses on the studies that have discussed the enzymatic effect on degradation. Also, the release behavior of starch-based DDSs was discussed by the distribution of enzymes in the human body.

4.3.1 Enzymes in the GIT

When starch passes through the GIT, most fractions of starch can be hydrolyzed in the small intestine by the action of pancreatic enzymes (Heinemann et al. 2005; Larionov et al. 1999; Putseys et al. 2010). The starch that escapes the digestion of pancreatic enzymes, namely, RS, can also be hydrolyzed by colonic microbiota (Yang et al. 2002). Thus, for oral administration, it is of high interest to assess the effect of the enzymatic degradation process on the drug release behavior since the enzymatic degradation of starch-based DDSs is inevitable.

4.3.1.1 Pancreatic Amylase in the Intestine

When to administrate by the oral route, due to the susceptibility of starch to the pancreatic enzyme, the drug release of starch-based carriers is highly affected by the enzymatic digestibility. The modification of starch is needed to increase the enzymatic resistance to some extent to avoid fast degradation of starch, thus protecting bioactive compounds from quick or unwanted release and achieving site-specific drug release. While at the same time, the capacity of enzymatic degradation should be retained, which can be used to trigger drug release at the targeted site or to accelerate drug release to achieve better controlled-release behavior (Thombre et al. 2009). Numerous modification methods, such as esterification (Bie et al. 2016; Situ et al.

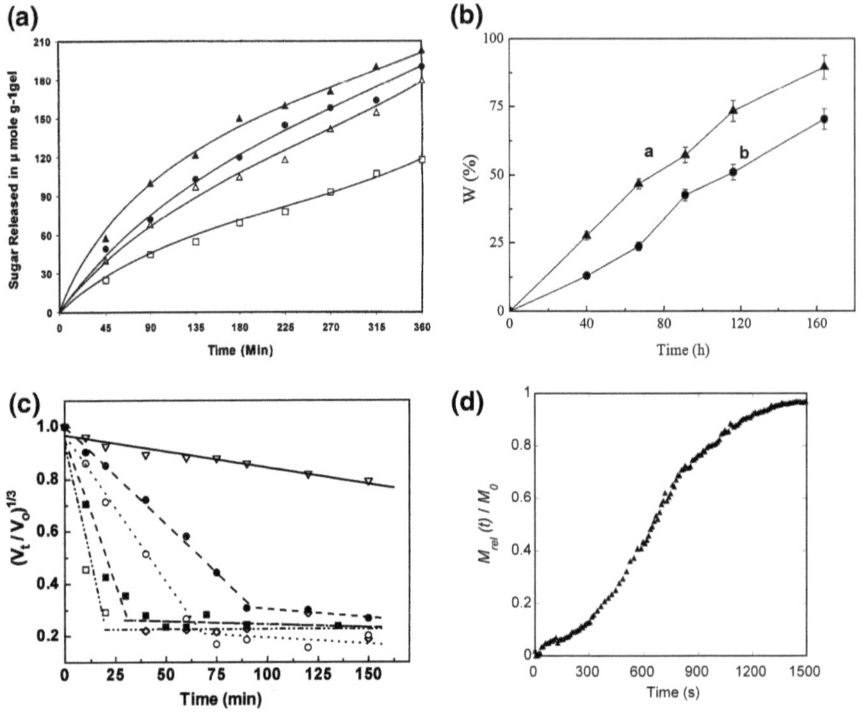

Fig. 4.6 Enzymatic degradation profiles of starch-based drug delivery. **a** Amount of sugar released as a function of time from the degrading hydrogel in the phosphate buffer of pH 6.8 with enzyme concentration 3.5 (□), 5.0(△), 6.5 (●), and 8.0 (▲) IU/mL at the physiological temperature 37 °C. Reprinted from Bajpai and Saxena (2006), Copyright (2006), with permission from Elsevier. **b** Degradation test of cationic starch nanospheres with α-amylase whose concentration (200 U/L) is close to that in serum (40–180 U/L) (a, with α-amylase; b, without α-amylase). Reprinted from Huang et al. (2013), Copyright (2013), with permission from Elsevier. **c** Linearized volumetric degradation profiles of starch microspheres as a function of a-amylase concentration: (▽) 5 IU/L, (●) 15 IU/L, (○) 25 IU/L, (■) 50 IU/L, and (□) 100 IU/L. Reprinted from Hamdi et al. (1998), Copyright (1998), with permission from Elsevier. **d** Fraction of protein released Mrel(t)/M0 as a function of time during enzymatic degradation of the gel particles by a-amylase. Reprinted from Li et al. (2011b), Copyright (2011), with permission from the Royal Society of Chemistry

2015) and recrystallization (Yoon et al. 2009), have successfully been used to modify starch to increase its enzymatic resistance (Table 4.2). Among those modifications, what is interesting is the natural and spontaneous tendency of amylose to form single helical molecular inclusion complexes with various ligands, termed V-amylose. Due to the high stability of the inclusion complexes during ambient storage, the advantages of these complexes include high yield (> 57%) and high drug loading (up to 68.16%) (Lesmes et al. 2008). The utilization of V-amylose complexes for the encapsulation of different biomolecules is highlighted here.

Different starch structures may result from the addition of ligands into starch pastes, such as the amorphous or disordered structure, polymorphs A, B, and C crystalline structures (highly ordered double helices of starch molecules) (Sarko and Wu 1978) and V-amylose (Carbinatto et al. 2016; Kim and Huber 2016; Lesmes et al. 2008). For example, when ethanol comprising varying β-carotene contents was added dropwise into starch pastes, no evidence was observed for self-association of A-, B-, or C-type crystalline structures or amorphous matrices (Kim and Huber 2016). Nonetheless, V-amylose structures (single helical structures of amylose stabilized by water molecules and ethanol) and starch-β-carotene V-amylose ordered structures were formed during the composite formation (Kim and Huber 2016). The changes of the starch structure by the addition of ligands (amorphous and V-amylose structures) may lead to varied enzymatic resistance and drug release behaviors. V-amylose has been associated with the formation of Type II (single helix) complex, which provides higher enzymatic resistance. The release of the ligands occurred by enzymatic hydrolysis of the complexes in two steps: initially, a rapid hydrolysis of amorphous regions of the complexes (Galloway et al. 1989; Jane and Robyt 1984; Kim and Huber 2016), and subsequently, a slow degradation of the highly ordered structure of V-amylose of them (Godet et al. 1996; Kim and Huber 2016). Specifically, in the case of starch-β-carotene composites, approximately 80–92% of the starch complexes were rapidly hydrolyzed within the first 20 min of in vitro digestion due to the amorphous structures, resulting in a rapid release of β-carotene (approximately 21–37%) (Kim and Huber 2016). On the other hand, other starch fractions (approximately 2–7%) completely escaped the digestion during the 120 min digestion period and retained β-carotene (approximately 52–66%) within the composites after 120 min of the in vitro digestion.

V-amylose inclusion complexes are believed to be comprised of packed single helices with a central hydrophobic cavity interconnected by amorphous regions of the polysaccharide chains, which render the complexes relatively stable in SGF and SIF without the presence of pancreatin (Carbinatto et al. 2016; Kim and Huber 2016; Lesmes et al. 2008; Yang et al. 2013). The presence of pancreatin triggered a significant acceleration of the release rate of drugs, due to the hydrolysis behaviors of the inclusion complexes by amylase. To form V-amylose inclusion complexes, the amylose content of starch is crucial because of the high tendency of amylose to form inclusion complexes. When cornstarch and HACS were used, the V-amylose starch–stearic acid complexes structure was significant; however, when waxy starch was used, there was no evidence of the formation of V-amylose complexes (Lesmes et al. 2008). Compared with the controls, enzymatic digestion significantly increased the release of stearic acid for HACS and cornstarch; however, no significant differences were noted in the case of waxy corn starch, suggesting waxy corn starch is not suitable for such a purpose of controlled release (Lesmes et al. 2008). Besides, the optimized proportion of ligands to starch is often needed to achieve a desired drug release behavior of drug–amylose inclusion complexes. The content of the V-amylose ordered structure can be increased by the increased amount of ligands incorporated into the composite. However, too many ligands increase the fraction of physical blending of ligands with amorphous matrices, leading to a fast drug release (Kim and Huber 2016).

4.3.1.2 Microbial Enzymes in the Colon

Among different approaches to achieve colon-specific DDSs, the use of polymers that can be specifically biodegraded by colonic bacterial enzymes holds assurance (Krishnaiah et al. 2002; Sinha and Kumria 2001). To reach the colon, starch should escape digestion in the saliva, the small intestine and reach the large bowel. Upon reaching the colon, the starch derivatives undergo assimilation by microorganisms, degradation by enzymes, and/or breakdown of the polymer backbone. Hence, the starch-based carrier will lose its mechanical strength, and the disintegration of the matrix will trigger a faster drug release.

The enzymatic resistance of starch-based carriers must be high enough to minimize the drug release in the upper GIT (Patten et al. 2009). Chemically modified starches, such as cross-linked starch and starch acetate (SA), have been applied to colon-specific DDSs in various dosage forms including MPs and NPs (Table 4.2). The enzymatic resistance was usually increased with the degree of modification (e.g., the cross-linked degree of cross-linked starch, and the DS of SA), which can decrease the drug release in the upper GIT (Bie et al. 2016; Situ et al. 2015; Thombre et al. 2009). For example, when the DS of SA reached 1.94, the RS content was 93.85%. In this case, within the first 8 h in SGF and SIF, the cumulative release percentage of drugs from all SA film-coated MPs was less than 16.5%. Only after transferred into the simulated colon fluid (SCF), the cumulative release percentage was then increased significantly, suggesting SA film-coated MPs have good colon-targeting performances (Bie et al. 2016). Therefore, it is an important way to adjust the degree of modification of starch to reach colon-specific delivery.

Blends of starch derivatives with other polymers that are resistant to digestion in the small intestine, such as pectin, chitosan, and alginate, can be an alternative for designing colon-specific DDSs (Table 4.2). For example, HACS is not digestible by pancreatic enzymes (Noakes et al. 1996) and has been used to be blended with casein to avoid degradation in the small intestine and minimize the drug release in the upper GIT (Patten et al. 2009). Besides, recently, physical modifications combining with enzymatic methods have been used to enhance the enzymatic resistance of modified starch for reaching the colon. RS-rich fractions can be obtained by hydrothermal treatment and retrogradation of native HACS (Thompson 2000). RS3, which was obtained by a high-temperature–pressure reaction, enzymatic debranching, and retrogradation process, resulted in a dramatic increase in the resistibility against enzymatic digestion (Situ et al. 2014). These RS3 film-coated MPs were found with no obvious change after 2 h in SGF. After immersed in SIF for another 6 h, only a few small holes were observed, while the MPs still kept its integrity. When immersed in the SCF for another 16 h, those MPs showed more holes and cracks emerged, and the integrity of the film was tampered (Situ et al. 2014). Accordingly, the release percentage of drugs was up to 25.31% within the first 8 h in the upper GIT and then reached 80.66% at 30 h in the colon (Situ et al. 2014).

In some cases, raw starch itself can have particular properties and enzymatic resistance without modification. Regarding this, Assam Bora Rice starch is one example. It has high amylopectin content (> 95% of amylopectin) and high crystallinity, giving it physical stability and enzymatic resistance in the GIT (Ahmad and Ashokanshu 2009). When using MPs made from this starch, insignificant release of drugs was found in the physiological conditions of the stomach and small intestine, and major drug release occurred in the caecal content (Ahmad et al. 2012a, b, 2013). Also, results showed that these drug-loaded MPs were distributed primarily to the lower GIT, whereas raw drug distributed predominantly in the upper GIT (Ahmad et al. 2012a, b, 2013). Future research may need to focus on the study of raw starches with particular features, which can address the challenges with DDSs but meanwhile avoid modification to reduce costs and potential negative health impact.

4.3.2 Human Serum Endoamylase

For injection administration, starch-based particles are needed to be transported through the blood circulation to the remote body sites to increase the drug concentration in tumor tissues during the treatment of localized tumors. Considering this, the biocompatibility and biodegradability of starch-based carriers are essential. Results showed that starch-based carriers could be prepared in a well-controlled way to achieve biodegradation without toxicity to human cells. One example is that 22% of ultrasmall (average diameter of 150 nm) anionic starch NPs, which functioned to retard the drug washout from the kidney microcirculation mechanically, was degraded after 4 days and did not present obvious cytotoxicity to the cells after 48 h (Huang et al. 2013).

In the case of non-intracellular drug delivery, the biodegradation of starch MPs and NPs in the blood circulation system is critical for drug delivery and clinical application, making these particles potential drug carriers. Degradable starch microspheres (DSMs) are one of the most outstanding examples to be applied in the blood circulation to increase the drug concentration at the localized tumor (Table 4.3). DSMs are cross-linked starch microspheres that can be degraded, with the half-time for a complete dissolution between 15 and 30 min in vitro (Ensminger et al. 1985). Drugs administered with DSMs will be retained in the capillary vessels for a long time, which allows increased extraction of the drugs. For example, because of the slow occlusion of DSMs, the drug was retained in the tumor at a concentration up to 11 times, with the biological half-time 13 times longer, than the drug was used alone without DSMs (Lorelius et al. 1984). The degree of drug retention was inversely related to the blood flow, with the retention increased as the blood flow decreased. The complete degradation of the DSMs by endogenous amylase and the reestablishment of normal blood flow could be achieved within 1 h after injection (Björses et al. 2011; Carter et al. 1992; Dakhil et al. 1982; Kovacs and Turowski 2002; Lorelius et al. 1984; Shiba et al. 2006; Sigurdson et al. 1986; Thom et al. 1988; Tuma et al. 1982; Yoshikawa et al. 1994). Because of this short lifetime, DSMs allow repeated treatment through "intra-arterial injection". In addition, the degradation degree of starch-based carriers can be modulated to acquire better drug delivery behavior. For example, when SA film-coated MPs were used, the drug release profile from the SA

Table 4.3 Effects of degradable starch microspheres (DSMs) injected into blood serum to enhance drug absorption

Therapeutics	Comments	References
Blood flow		
Actinomycin D	Blood flow of the renal artery of dogs decreased to less than 5% of the control flow	Tuma et al. (1982)
Superoxide dismutase and catalase	Blood flow decreased rapidly immediately after the DSMs treatment and returned to the original level within 40 min	Yoshikawa et al. (1994)
Drug retention and concentrations		
Angiotensin II	12-fold improvement of the retention time in liver and tumor versus raw drugs	Yoshikawa et al. (1994)
Actinomycin D	15-fold higher retention in the cortex and 13-fold higher retention in the medulla versus raw drug	Carter et al. (1992)
Effect of treatment		
Gene (AxCALacZ)	β-galactosidase expression was greater and more selective in tumors than the carrier alone	Shiba et al. (2006)
Superoxide dismutase and catalase	Five repeated DSM treatments led to a significant reduction in tumor size	Yoshikawa et al. (1994)
Angiotensin II	A significantly improved tumor to liver ratio (5.22:1) in combination treated animals	Carter et al. (1992)
Hepatic arterial bischlorethylnitrosourea	10 mL of microspheres injected into the hepatic artery to transiently (for 15–30 min) reduced hepatic row by 80–100% in the five patients	Dakhil et al. (1982)

DS 2.6 film in a pH 7.4 buffer solution (5-days incubation) showed discontinuities (Tuovinen et al. 2004a). However, the release of drugs was continuous in human serum in vitro during a 7-day incubation, i.e., enzymes enhanced the drug release. Based on that, α-amylase was incorporated into the SA films to enhance the drug release from the films (Tuovinen et al. 2004a).

4.3.3 Cellular Compartment Enzymes

Intracellular delivery of drugs may enable stronger, more selective, and prolonged pharmacological actions and thereby reduce the side effects. Starch MPs and NPs can be used for intracellular drug delivery, which is often achieved by injection. Cytosol has lots of hydrolases inside, especially lysosomal hydrolases (Jovic et al. 2010). Besides, the enhanced expression of enzymes can be induced by the diseased microenvironment of cells (Yang et al. 2011). All of these form the base of enzyme triggers of starch-based DDSs for intracellular delivery.

In recent years, numerous efforts have been dedicated to designing smart carriers with well-controlled enzymatically responsive drug release behavior. One typical and prominent example is starch-decorated gated mesoporous materials. Recently, gated mesoporous materials have become appealing since these materials contain switchable molecular-based entities that control the on-command release of previously entrapped guests. One way to achieve this switchable release is to anchor starch derivatives to the surface of gated mesoporous silica materials, which can respond to the enzyme at tumor sites (Bernardos 2010) or intestinal fluids (Acosta et al. 2014). The starch derivatives can inhibit drug delivery due to the formation, around the pore outlets, of a dense monolayer of starch molecules. When encountering enzymes, the starch was hydrolyzed with the subsequent uncapping of the pores and drug release (Fig. 4.7a, b). The drug release behavior showed almost the "zero release" obtained in the absence of enzyme, but displayed drug release in the presence of the corresponding enzyme (Fig. 4.7c). The less hydrolysis of starch, the lower was the delivery rate (Acosta et al. 2014). Moreover, results suggest that those starch-gated silica mesoporous particles can devoid of unspecific cell toxicity due to endocytosis targeting them at the autolysosomes (Bernardos 2010).

Drugs for the treatment of cancer and other diseases in starch-based DDSs can be divided into drugs, peptides, DNA or small interfering RNA for gene treatment (Table 4.2). Among those bioactives, starch has a high potential to overcome the shortcomings of conventional polymer carriers of gene delivery. These commonly existing polymer carriers usually have high cytotoxicity caused by high cationic charge density, as well as nonbiodegradability and poor cell biocompatibility (Nitta and Numata 2013). Hence, the gene delivery application of starch-based nanocarriers is emphasized here. Cationic starch derivatives are highly needed to form DNA complexes through electrostatic interaction with negative DNA or to interact with negatively charged cell membranes. The cationic surface of starch functionalized with a quaternary amine (Constantin et al. 2003; Engelberth et al. 2015; Sieradzki et al. 2014) or ethylenediamine groups (Thiele et al. 2017) are among the recently published strategies to use starch-based excipients for nucleotide delivery. Besides, another way to prepare cationic starch derivatives is conjugating starch with existing classical gene delivery polymers, such as PEI, which is the mostly used cationic polymer to deliver genes but with high cytotoxicity (Godbey et al. 1999). Starch (Yamada et al. 2014) or hydroxyethyl starch (HES) (Noga et al. 2013, 2014) have been conjugated with 22 kDa PEI to reduce the cytotoxicity of PEI while remaining

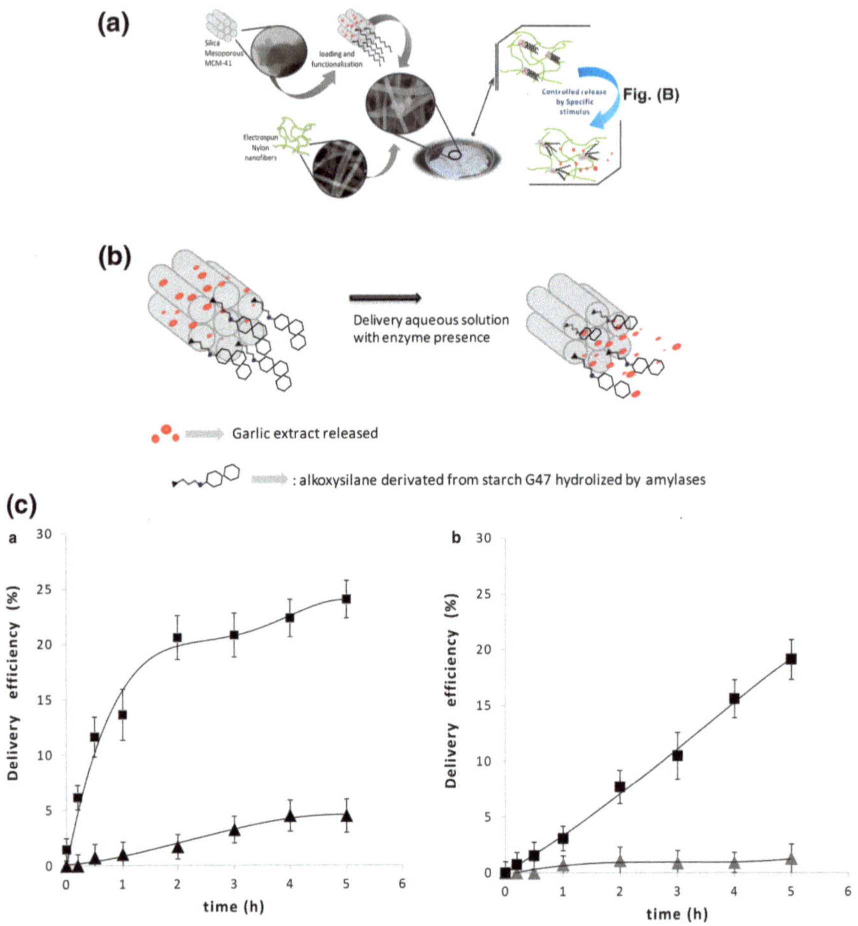

Fig. 4.7 Characterization of the release behavior of mesoporous silica microparticles (MCM-41) solids functionalized in the outer surface with hydrolyzed starch, acting as molecular gates. **a** Those particles were adsorbed onto electrospun nylon-6 nanofibrous membranes to form composite materials. **b** Mechanism of enzyme-induced release of the entrapped cargo. **c** Delivery of garlic extract from enzyme-response materials: (a) starch-gated MCM-41 solid particles and (b) starch-gated MCM-41 solid/electrospun nylon-6 nanofibrous membranes composite. Release in the presence of pancreatin (■) and release in the enzyme absence (▲). Reprinted from (Bernardos 2010). Copyright (2014), with permission from the American Chemical Society

the high transfection efficiency. To achieve this goal, the shielding effect, manifested by the reduced surface charge, is needed for starch-based gene DDSs. However, a low surface charge of carriers leads to comparatively low gene transfer efficiency (Kichler 2004). Concerning this, a good controlled shielding and deshielding is a must for starch-decorated gene nanocarriers, which can be achieved by enzymatically catalyzed hydrolysis. For instance, HES-mediated particles showed the reduced in vitro transfection in the absence of α-amylase, due to the excellent shielding effect of HES, while the addition of amylase caused an increase of up to three orders of magnitude in the gene transfer by these HES–PEI copolymers (Noga et al. 2013).

Gene delivery systems have to overcome several systematic and cellular barriers to deliver genes to the nuclear of diseased cells. All of these barriers may pose different challenges to the gene delivery system thus affect the gene transfection efficiency. First, cationic starch/DNA complexes need to be stable in blood serum to avoid DNA enzymatic degradation. Results confirmed cationic starch derivatives could shield DNA from degradation, allowing intact DNA to be released in the cells (Noga et al. 2013, 2014; Sieradzki et al. 2014). After the complex enters the cell, it must escape from the lysosome, which is the second barrier, for DNA to reach the cell nucleus. Cationic starch derivatives need to have the ability of lysosomal escape, which can be achieved by functionalizing with a quaternized structure (Constantin et al. 2003; Engelberth et al. 2015; Sieradzki et al. 2014) or conjugating with PEI (Noga et al. 2013, 2014; Yamada et al. 2014). The third atom of amino nitrogen of the structure of cationic quaternized starch is a protonable atom, which makes the polymeric network an effective "proton sponge" at virtually any pH (Kichler 2004). This may permit lysosomal escape and thus prevent DNA from the lysosomal degradation. The subsequent barrier could be the decomplexation between the carrier and DNA, and the translocation of the genetic drug into the cell nucleus (Lechardeur and Lukacs 2002; Lechardeur et al. 1999). The degradation of starch is crucial to this decomplexation process and the acceleration of the DNA release for improving the transfection efficiency (Engelberth et al. 2015; Noga et al. 2013, 2014; Thiele et al. 2017; Yamada et al. 2014). This was attested by the cationic starch carrier quaternized by 3-chloro-2-hydroxypropyltrimethylammonium chloride (Sieradzki et al. 2014). Regarding this, it was turned out that the membrane barrier (i.e., the cellular uptake) and lysosomal escape barrier were probably not the limiting factors for efficient transfection, while decomplexation process by the amylase degradation of starch may be a major factor for efficient transfection (Sieradzki et al. 2014).

Gene transfection is a result of complex factors, such as the particle size, the polymer M, the length of side chains, and the charge spacing along the polymer backbone (Wolfert et al. 1999). For starch-based gene delivery carriers, the whole panorama of the factors and their effects on gene delivery are still not well elaborated. For example, the optimized parameters, such as the DS of starch (Yamada et al. 2014), nitrogen/phosphate atom (N/P) ratios (Sieradzki et al. 2014; Yamada et al. 2014), and the M of starch (Yamada et al. 2014), have been tested to acquire high transfection efficiency. However, for starch-*g*-PEI and HES-*g*-PEI, the higher M of starch or HES led to a greater degree of transfection (Holmberg et al. 1994; Noga et al. 2013, 2014). Those factors interact with each other to affect gene transfection efficiency, e.g., the N/P ratios can be modulated by the DS and *M* of starch (Sieradzki et al. 2014; Thiele et al. 2017). Moreover, the shielding efficiency of cationic starch derivatives in blood serum can be influenced by not only the enzymatic degradation but also the ionic strength of the medium. An increase of NaCl concentration was found to increase the amount of DNA released from MPs, which were prepared by cross-linked starch followed by amination (Constantin et al. 2003). More work needs to be undertaken to find out the structure–functionality relationship and to improve the delivery efficiency of cationic starch/DNA complexes.

Owing to the different functions that enzymes have in different biological processes, together with amylase, disease-associated enzyme dysregulations have become an emerging target for medications. Ester bonds are often incorporated for targeting phosphatases, intracellular acid hydrolases, and several other esterases. Take HES microcapsules as an example, after 20% BSA was released from HES microcapsules within 30 min, 80% remained entrapped in the microcapsules for 4 days. Total amounts of BSA were released only after the complete degradation of HES microcapsules by esterase and amylases (Devy et al. 2006). First, esterase catalyzed the cleavage of the ester linkage, and then amylases further hydrolyzed the remaining polymers to glucose (Tuovinen et al. 2004b). Besides, in the case of carriers prepared from starch, HA, and its salts, both hyaluronidase (HAase) and amylases in the lysosome at the tumor can be utilized to trigger drug release, since tumor sites often have a higher content of HAase than normal tissues (Kwag et al. 2014; Li et al. 2015). For example, to reduce the toxicity of drugs, the separate release of two drugs could be achieved through separately enzymatic digestion of carriers in a complicated nanocarrier (Fig. 4.8a). This mulberry-like complicated nanocarrier was assembled with positively charged apogossypolone loaded starch micelles and negatively charged DOX-loaded HA NPs via electrostatic interaction (Fig. 4.8b) (Li et al. 2015). Almost no apogossypolone release was found within 80 h, and less than 30% of DOX was released into the outer buffer solutions (pH 7.4) even after 72 h (Li et al. 2015). When HAase was added, more than 80% of DOX was released within 10 h, due to the degradation of the HA layer (Fig. 4.8c). Then, apogossypolone was released after 80 h due to the enzymatic degradation of amphiphilic starch micelles. These nanocarriers exhibited significantly higher antitumor efficiency compared with the free drugs combination or single drug-loaded NPs.

4.4 Temperature

Combinations of thermosensitivity with starch-based carriers have some special advantages, which can be applied in different ways. First, it is possible to design a gelation system to be active only at body temperatures. As a result, the carrier is in the liquid form under laboratory conditions but after injection into the body, it is converted to the gel form (Dong 2016; Li et al. 2014). Moreover, the temperatures sensitivity of the carrier can enhance the drug release at the normal temperature of the body. For example, because the glass transition temperature (25 °C) of the

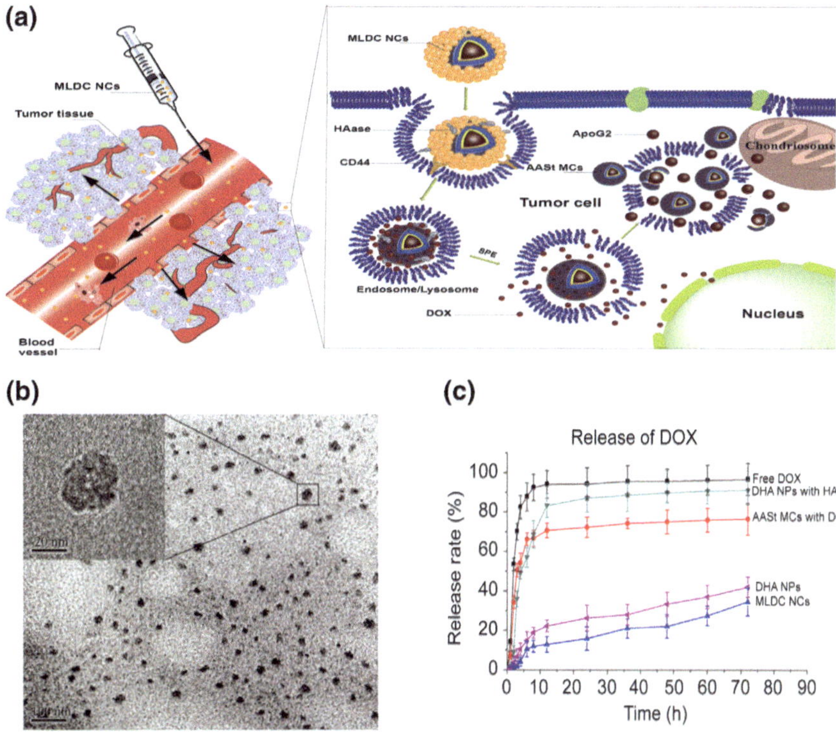

Fig. 4.8 Characterization of the drug release behavior of mulberry-like dual-drug complicated nanocarriers (MLDC NCs). **a** Schematic representation of drug delivery and intracellular release process of MLDC NCs. **b** TEM images of MLDC NCs. **c** In vitro release profile of doxorubicin (DOX) from MLDC NCs at 37 °C, free DOX, DOX-loaded amphiphilic starch micelles (AASt MCs), and hyaluronic acid nanoparticles (DHA NPs). DHA NPs with hyaluronidase (HAase) (10 U/mL) were used as the control. Reprinted from Li et al. (2015), Copyright (2015), with permission from Elsevier

plasticized starch was lower than the human body temperature, the relaxation of macromolecular chains of starch polymers favored the swelling of carriers and the diffusion of drugs to the external environment (Schmitt et al. 2015). Furthermore, if this sensitivity is designed for higher temperature, it would be possible to apply external heating to the local area like a cancer site, leading to the release of the drug in the specific location. Tumor cells seem to be more sensitive to heat-induced damage than normal cells (Ganta et al. 2008). To date, the methods used to synthesize thermosensitive starch derivatives can be classified into two broad categories: (a) the incorporation of thermosensitive polymer units into starch, and (b) the preparation of starch derivatives with the hydrophobic–hydrophilic balance.

The incorporation of thermosensitive polymer units into starch is an easy way to endow starch with thermoresponsiveness. In turn, the addition of starch will enhance biodegradation of its composite polymers. Grafting PNIPAM and its derivatives onto starch is the most investigated approach to prepare starch-based thermosensitive polymers (Bakrudeen et al. 2016), and these polymers show good thermoresponsiveness to the human physiological/pathological environment. These starch-based carries can not only be applied for hyperthermia but also developed a background of "smart" implantable DDSs for hypothermia cases. For example, copolymers of poly(*N*-isopropylacrylamide-*co*-*N*,*N*-dimethylacrylamide (DMAAm)) (PNIPAM-*co*-DMAAm) were grafted onto starch MPs with an LCST of around 36 °C, and then encapsulated in cellulose acetate butyrate microcapsules for the temperature-controlled release of drugs (Fundueanu et al. 2010). When the environment temperature is around the LCST, a phase transition leading to the swelling or shrinkage of MPs occurred. Thus, a significant amount of bioactive compounds was released when the temperature of the human body decreased below the normal temperature. Nonetheless, when the temperature was high than the volume phase transition temperature (VPTT) (38 °C), due to the diffusion of the drug facilitated by the higher temperature, the release was fast, even if the microspheres were in the shrunken state (Fundueanu et al. 2010).

The thermoresponsiveness of starch-based polymers can also be obtained by controlling the hydrophobic–hydrophilic balance of starch polymeric chains. Even though starch is insoluble in cold water, modification by grafting an appropriate amount of hydrophobic groups onto starch chains can generate starch-based water-soluble polymers by disrupting the inter- and intramolecular hydrogen bonding of starch (Funke and Lindhauer 2001). For example, 2-hydroxy-3-butoxypropyl starch (LCSTs: 4.5–32.5 °C) (Ju et al. 2012) and 3-[2-butoxy(ethoxy)m]-2-hydroxypropyl starch ethers (LCSTs: 17.5 to 55.0 °C) (Ju et al. 2014) were synthesized by introducing ethers to starch as hydrophobic reagents to form stable colloidal micelles. Compared with traditional polymeric micelles, the main advantage of these thermoresponsive polymeric micelles is that these micelles can achieve the on-off release of the incorporated drug in response to the temperature change. The drug release was accelerated dramatically above the LCST. For example, at 20 °C (below the LCST: 32.5 °C), about 38% of the drug was released from the 2-hydroxy-3-butoxypropyl starch micelles within about 100 h in water, with about 62% of the drug remaining in the core of the micelles due to the stable micelle structure. In contrast, at 40 °C (above the LCST), about 90% of the drug was released from the micelles within around 100 h (Ju et al. 2012). Increasing hydrophobicity usually lowers the LCST of amphiphilic starch ethers, making it easier to self-assemble into micelles. It was verified by the result that the LCST was decreased linearly with the increased MS of ethers due to the increased hydrophobic interactions between hydrophobic groups of micelle (Ju et al. 2012). High hydrophilicity of starch polymers will increase the LCST by increasing the difficulty of the dissociation of water molecules around the polymer chains during the phase transition (Ju et al. 2014).

4.5 Redox

Starch-based redox-sensitive materials can be easily achieved by coupling redox pair to starch. The mild reaction conditions of thiol–disulfide exchange render it an appealing approach to constructing disulfide-containing materials, which allowed the controlled release of encapsulated drugs from starch NPs (Meng et al. 2009) and micelles in the presence of disulfide-containing cross-linkers (Wu et al. 2016; Yang et al. 2014b; Zhang et al. 2013b), such as *N,N*-bisacryloylcystamine. The drug release from redox-responsive starch-based carriers was effectively hindered under normal physiological conditions, while was accelerated due to the cleavage of the disulfide bond in the core, caused by the presence of GSH or dithiothreitol (Yang et al. 2014b; Zhang et al. 2013b).

Another focus used for engineering starch-based redox-sensitive materials is diselenide linkage. There have been reports on the coupling reaction of methoxypolyoxyethylene amine (mPEG) with 3,3′-diselenodipropionic acid, with the subsequent grafting reaction with starch to prepare mPEG–SeSe–starch micelles (Chen et al. 2016a, b). Interestingly, the micelles of mPEG–SeSe–starch could be disrupted because of the cleavage of diselenide bonds not only by a high concentration of GSH in the cytoplasm but also by a low concentration of hydrogen peroxide (H_2O_2, 0.1%, v/v) (Fig. 4.9a). The size of micelles increased markedly under both conditions, and it was more significant in the case of GSH (Fig. 4.9b). These results suggest that the diselenide bonds can be considered as a novel dual redox/ROS-sensitive linkage, being responsive to reducing and oxidizing conditions. Moreover, compared with disulfide bonds, it was noticeable that DOX-loaded micelles with diselenide linkage showed higher activity and could be more quickly disrupted in a reducing environment and subsequently rapidly released drug (Fig. 4.9c). This may be due to the lower bond energy of diselenide than disulfide (Xu et al. 2013)—diselenide linkage was more sensitive than disulfide linkage at the same concentration of GSH. All of these results indicate that the diselenide linkage outperforms the disulfide linkage in the DDS application and become a high potential redox couple to endow materials with redox responsiveness. More work is needed to illuminate its properties and enlarge its applications in starch-based DDSs.

4.6 Magnetic Field

Besides endogenous environmental conditions, starch-based DDSs can be designed to be responding to the external stimuli, which can be controlled by artificial manual work. To date, the most studied external stimulus in starch-based DDSs is a magnetic field. Despite having no harm to human tissues, MNPs have an issue with their tendency to aggregate and sediment, which limits its application in DDSs. Coating its surface with starch derivatives, such as CMS (DS = 0.33) (Saboktakin et al. 2009b), can solve this problem and give good biocompatibility and biodegradation.

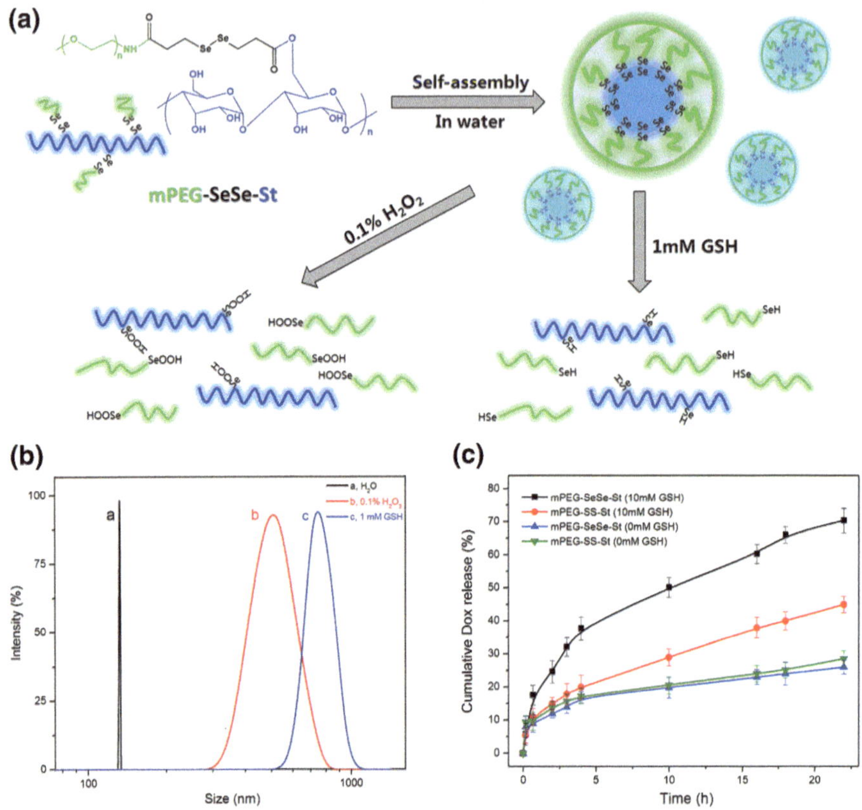

Fig. 4.9 Characterization of the release behavior of methoxypolyoxy-ethylene amine (mPEG)–SeSe–starch (St) micelles. **a** Schematic of the oxidation- and reduction-responsive transformation of mPEG–SeSe–St micelles. **b** Dynamic light scattering results of mPEG–SeSe–St polymer micelles in water (a), with H_2O_2 (b), or glutathione (GSH) (c) after 4 h. **c** GSH-triggered doxorubicin release profiles from mPEG–SeSe–St micelles and mPEG–SS–St micelles. Reprinted from Chen et al. (2016a), Copyright (2016), with permission of the Royal Society of Chemistry

Meanwhile, the addition of MNPs gives the starch magnetic-responsive properties and enhances its ability to target at diseased tissues and the bioavailability of drugs (Fig. 4.2). Research (Alexiou et al. 2001) showed that after injection of starch-coated MNPs, the tumor size decreased and there was a complete remission within 15 days.

Ferrogels made from MNPs embedded in starch gels have attracted attention due to the promise of combining a relatively safe physical stimulus with generally considered biocompatibility (Hamidian and Tavakoli 2016; Xuan Phuc et al. 2012). When the EMF was applied, there are mixed results of the effect of EMF on the drug release behavior. It was noted that the swelling capacity and the drug release was enhanced by the magnetic-responsive properties due to the improved polymer relaxation of starch polymers because of high-frequency magnetic pulses (Xuan Phuc et al. 2012). Usually, the higher the iron content contained in the MNPs, the higher is its magnetic responsivity. However, it is noteworthy that MNPs can chelate to several polymer chains, and thus they can act as a cross-linking agent for constructing MNPs-starch nanocomposite hydrogels. As a result, the introduction of high content Fe_3O_4 particles can inhibit the swelling capacity of nanocomposite hydrogels due to the interaction of Fe_3O_4 with the functional groups of starch (Guilherme et al. 2012). This was testified by the drug release behavior of starch-*g*-polyester nanocomposite hydrogel incorporated with Fe_3O_4 NPs (Hamidian and Tavakoli 2016). It was found that higher concentrations of the precursors $FeCl_2$ (0.05 M) and $FeCl_3$ (0.025 M) led to lower swelling, thus hindering the drug loading capacity, the drug release rate, and the cumulative drug released amount, owing to the chelation of Fe_3O_4 NPs with ester groups of starch polymers (Hamidian and Tavakoli 2016). Besides, the tortuosity effect, which is the result of the disposition of the mineral structures in the hydrogel, is another factor that may disturb the release of drugs (Guilherme et al. 2012). Therefore, considering the opposite effects of the introduction of MNPs on the swelling capacity and the drug release, an appropriate percentage of MNPs in starch-based ferrogels could be critical for tailoring the drug release performance.

4.7 Dual and Multi-stimuli

The human body is a complex collection of different environments. Dual and multi-responsive starch-based DDSs have been an interesting ongoing investigation for scientists, especially when materials that are responsive to a single stimulus may not be effective enough for in vivo therapy due to the physiological variations in some diseases (Fallingborg et al. 1993; Sasaki et al. 1997). Efforts so far have been focused on developing starch-based DDSs with multi-responsiveness to mainly pH, temperature, enzyme, and magnetic field. The combinations of two or more different stimuli can be rationally chosen to further increase the versatility and specificity of triggered DDSs. For example, using stimulus-responsive starch polymers as coatings for MNPs can selectively increase the drug concentrations in cancer cells, thus improving its targeting efficiency and greatly decreasing the side effects of drugs (Fig. 4.2). Moreover, a multi-responsive carrier may have better ability to sense very slight changes in the environment such as small variations in pH and temperature.

Among various stimuli, the combination of pH with other stimuli (e.g., temperature and magnetic field) is the most common approach to endow dual/multi-stimuli-responsive starch-based carriers that have been studied. As discussed in Sect. 4.2, the pH responsiveness can be achieved through introducing anionic groups (Bardajee and Hooshyar 2013; Fathi et al. 2015; Guilherme et al. 2012; Lima-Tenório et al. 2015) or PEC (Saikia et al. 2015), through drug–polymer interactions (Likhitkar and Bajpai 2012; Likhitkar and Bajpai 2014; Zohreh et al. 2016), or through pH-sensitive Schiff base bonds (Yang et al. 2009). These mechanisms can be combined to contribute to the pH sensitivity of multi-responsive starch-based DDSs. For example, for aminated starch/ZnO-coated MNPs (Saikia et al. 2017) and for PAMAM-modified Fe_3O_4 NPs coated by PEGylated starch-*co*-poly(acrylic acid) (Pourjavadi et al. 2015), the pH responsiveness was not only modulated by the interactions between the hydrophilic and ionic groups of starch polymers but also by the hydrogen bonding between the starch-coated NPs and the drug. The breakage of hydrogen bonding under acidic conditions allowed release of more drugs in a tumor environment (Pourjavadi et al. 2015; Saikia et al. 2017).

Moreover, new methods and mechanisms may account for the pH responsiveness. For example, contrary to the hydrolysis of disulfide bonds, the formation of intermolecular disulfide bonds at low pH was found in thiolated starch-coated MNPs (Saikia et al. 2014). As a result, higher amounts of the drug were released from the carrier at high pH compared to low pH, due to the electrostatic repulsion of thiol groups (–SH) of the polymer chains (Saikia et al. 2014). Besides, instead of grafting a negative carboxylic acid to starch, the amination of starch is an alternative way to provide starch-coated MNPs with pH sensitivity. Contrary to anionic carboxylic groups, in an acidic medium, the amine groups of the polymer chains were protonated, resulting in the polymers with a high positive charge density. This charge formation facilitated the repulsion between polymer chains and hence improved the swelling behavior, leading to a more drug release under acidic conditions and facilitating the anticancer drug to be delivered to tumor cells (Saikia et al. 2015, 2016, 2017).

Temperature is also among the most frequently studied stimulus in dual/multi-stimulus-responsive starch-based carriers. In particular, polymers displaying an LCST slightly above the body temperature are of particular interest for magnetically actuated drug delivery due to the promise of coupling drug delivery to hyperthermic cell ablation or to the induction of a heat-shock response. An example of efforts for this application is magnetic (PNIPAM-*co*-itaconic acid)-*g*-(starch MNPs) nanohydrogels. Incorporating more hydrophilic itaconic acid introduced pH sensitivity and relatively high LCST values of this starch derivative (Fathi et al. 2015). The LCST of (PNIPAM-*co*-itaconic acid)-*g*-(starch MNPs) was above 37 °C among pH values of 3, 4, and 8 (Pourjamal et al. 2016). By integration with thermoresponsive materials, the localized heating caused by magnetic particles under an alternating current EMF can be used to realize controlled drug release through local de-gelling process caused by the phase or volume transition of thermosensitive polymers.

It is already known that introducing thermosensitive groups to starch could permit the establishment of starch-based multi-responsive DDSs with thermosensitivity (Fathi et al. 2015). Besides PNIPAM, poly(*N*,*N*-dimethylaminoethyl methacrylate) (PDMAEM) is another polymer that can bring temperature responsiveness to drug delivery carriers. Interestingly, PDMAEM showed a pH-dependent LCST (Plamper et al. 2007). The pH sensitivity of this polymer is due to the protonation of the tertiary amine groups of PDMAEM in an acid medium (Han et al. 2013), with the resulting increased swelling capacity of this polymer matrix. Temperature/pH-responsive semi-IPNs hydrogels have been prepared by using PDMAEM as a matrix and native potato starch or anionic starch-*g*-PAN copolymer as the entrapping polymer (Dragan et al. 2016). The different nature of the entrapping polysaccharide in the hydrogel could result in different pH- and temperature-sensitive behaviors. Specifically, when anionic starch-*g*-PAN was used as the entrapping polymer, the degree of swelling was always lower than that in the case of potato starch at pH 3.0–8.0 (Dragan et al. 2016). This fact was attributed to the decreased hydrophilicity of the composite gel caused by the electrostatic interactions between protonated PDMAEM matrix and negative charges of starch-*g*-PAN. In addition, different entrapping polymers led to different VPTT. The VPTT of potato starch composite hydrogel was 39 °C, while the VPTT of the anionic starch-*g*-PAN composite gel was 37 °C (Dragan et al. 2016). Because of that, when the drug release was conducted at 37 °C, the drug release from potato starch composite hydrogels was slightly faster than that from starch-*g*-PAN (Dragan et al. 2016). This study could be a good example to illustrate that multiple factors can contribute the drug release behavior in a more sophisticated way.

It is notable that the change in drug–polymer interactions can also be considered to be a thermosensitivity mechanism for starch-based pH/temperature- or pH/temperature/magnetic field-responsive DDSs (Likhitkar and Bajpai 2012; Likhitkar and Bajpai 2014). An example of these DDSs was pH/temperature-sensitive hydrogels prepared by grafting copolymerization of L-aspartic acid with starch. In this system, the temperature change led to the association and/or dissociation of hydrogen bonding between the carboxylic groups of grafted starch polymers and amide groups of the drug (fasudil), showing thermosensitivity of this DDS (Vakili and Rahneshin 2013). Besides, due to the carboxylic groups, all of the hydrogels exhibited a high swelling ratio; in aqueous media, this value was greater at higher pH than at lower pH.

The magnetic responsiveness of starch-based multi-responsive DDSs that have been reported most widely usually combines with the pH and/or temperature responsiveness of the polymer. In these studies, the pH- and thermosensitive properties have been extensively studied. However, only a few studies have focused on the effect of EMF on the drug release of those carriers. Even so, there were discrepancies among the results, which can also be seen in the case of ferrogels in Sect. 4.6. For example, when starch-based superparamagnetic NPs were exposed to a high magnetic field, an enhanced diffusion process and the easier release of the drugs occurred because of the enlarged porous nanostructure channels produced by the EMF (Likhitkar and Bajpai 2014). Another example is superabsorbent starch-*co*-DMAAm-*co*-AA hydrogel composites, no change in the swelling performance was found with the use of EMF. The drug release was increased with the increased content of Fe_3O_4 in the composite, not because of higher magnetic responsivity, but due to the decreased affinity of the solute (albumin) with the hydrogel network (Guilherme et al. 2012). However, in the case of MNPs-PAA-*g*-starch hydrogels, EMF resulted in a lower swelling capacity and thus less drug release (Bardajee and Hooshyar 2013). This can be explained by the close-packed shell of MNPs, which can block the diffusion of the drug out of the MNPs-PAA-*g*-starch hydrogel. This close-packed shell structure was a result of the reduced distance between MNPs caused by the attraction force around neighboring MNPs induced by a given EMF (Bardajee and Hooshyar 2013). These results are even controversial with each other. Yet, there is still no trends on the effect of EMF on the drug release behavior, which may be due to other factors such as polymer properties, or the combined effect of all factors. More research is needed regarding this matter.

In some cases, enzymatic degradation of starch-based multi-responsive DDSs could be addressed to acquire the dedicated control of the drug release. One typical example of this is to achieve zero-order release of drugs, which is desired from a pharmaceutical point of view. Although many starch-based hydrogels were reported to be successful for the delivery of drugs to the colon (Abdel Ghaffar et al. 2016; Bajpai and Saxena 2004b; Desai 2005, 2007; Liu et al. 2011; Saboktakin et al. 2011; Tan et al. 2016), most of them could not give zero-order release in the colon. To achieve that, the degree of cross-linking of hydrogels should be low (Peppas 1997). However, hydrogels with a low cross-linking degree may release appreciable amounts of the drugs at low pH. By taking advantage of pH changes and enzymatic digestion along the GIT, the zero-order kinetics of drug release of starch-*g*-PAA semi-IPNs hydrogels can be achieved (Bajpai and Saxena 2004a, b). Another example is starch functionalized graphene nanosheets (Liu et al. 2015). The pH-sensitive drug release of this carrier was not only due to the partial dissociation of hydrogen bonding between the drug and starch–graphene nanosheets under acidic condition, but also attributed to the degradation of starch by amylase. With the double actions of acid microenvironment and diastase in cancer cells, the drugs experienced a pH-sensitive and starch-mediated sustained release process that had potential advantages for improving therapeutic efficacy (Liu et al. 2015).

So far, many efforts have been dedicated to preparing more functional starch-based multi-responsive DDSs to widen its application for medicinal treatment. Glucose and pH-responsive starch-based DDSs is an effort for glucose-responsive "closed-loop" medications that mimic the function of the healthy pancreas and work in a self-regulated manner (Veiseh et al. 2015). Glucose and pH-responsive luminescent starch have been facilely fabricated via mixing the aldehyde-containing, aggregation-induced dye with CMS and 3-aminophenylboronic acid in a one-pot procedure, in which 3-aminophenylboronic acid can serve as the linkage for conjugation of the dye and CMS (Liu et al. 2016). The Schiff base bonds between the aldehyde group of the emission dye and the amino group of 3-aminophenylboronic acid provided the luminescent starch with pH responsiveness, while the reaction of the phenyl borate group of 3-aminophenylboronic acid with the hydroxyl groups of CMS gave the luminescent starch glucose-responsive properties (Liu et al. 2016). These dynamic bonds can respond to both pH and glucose, which are very useful for the fabrication of responsive drug delivery carriers.

References

Abdel Ghaffar AM, Radwan RR, Ali HE (2016) Radiation synthesis of Poly(Starch/Acrylic acid) pH sensitive hydrogel for rutin controlled release. Int J Biol Macromol 92:957–964. https://doi.org/10.1016/j.ijbiomac.2016.07.079

Abhari N, Madadlou A, Dini A, Hosseini naveh O (2017) Textural and cargo release attributes of trisodium citrate cross-linked starch hydrogel. Food Chem 214:16–24. https://doi.org/10.1016/j.foodchem.2016.07.042

Acosta C et al (2014) Polymer composites containing gated mesoporous materials for on-command controlled release. ACS Appl Mater Interfaces 6:6453–6460. https://doi.org/10.1021/am405939y
Ahmad M, Ashokanshu B (2009) Isolation and physicochemical characterization of Bora rice starch from Assam as pharmaceutical excipients. J Pharm Res 2:1299–1303
Ahmad MZ, Akhter S, Ahmad I, Singh A, Anwar M, Shamim M, Ahmad FJ (2012a) In vitro and in vivo evaluation of Assam Bora rice starch-based bioadhesive microsphere as a drug carrier for colon targeting. Expert Opin Drug Delivery 9:141–149. https://doi.org/10.1517/17425247.2012.633507
Ahmad MZ, Akhter S, Anwar M, Ahmad FJ (2012b) Assam Bora Rice starch based biocompatible mucoadhesive microsphere for targeted delivery of 5-fluorouracil in colorectal cancer. Mol Pharm 9:2986–2994. https://doi.org/10.1021/mp300289y
Ahmad MZ, Akhter S, Anwar M, Kumar A, Rahman M, Talasaz AH, Ahmad FJ (2013) Colorectal cancer targeted Irinotecan-Assam Bora rice starch based microspheres: a mechanistic, pharmacokinetic and biochemical investigation. Drug Dev Ind Pharm 39:1936–1943. https://doi.org/10.3109/03639045.2012.719906
Alexiou C et al (2001) Magnetic mitoxantrone nanoparticle detection by histology, X-ray and MRI after magnetic tumor targeting. J Magn Magn Mater 225:187–193. https://doi.org/10.1016/s0304-8853(00)01256-7
Allen TM, Cullis PR (2004) Drug delivery systems: entering the mainstream. Science 303:1818–1822
Athira GK, Jyothi AN (2015) Cassava starch-poly (vinyl alcohol) nanocomposites for the controlled delivery of curcumin in cancer prevention and treatment. Starch - Stärke 67:549–558
Bajpai SK, Saxena S (2004a) Dynamic release of riboflavin from a starch-based semi IPN via partial enzymatic degradation: part II. React Funct Polym 61:115–129. https://doi.org/10.1016/j.reactfunctpolym.2004.04.004
Bajpai SK, Saxena S (2004b) Enzymatically degradable and pH-sensitive hydrogels for colon-targeted oral drug delivery. I. Synthesis and characterization. J Appl Polym Sci 92:3630–3643. https://doi.org/10.1002/app.20283
Bajpai SK, Saxena S (2006) Flow through diffusion cell method: a novel approach to study in vitro enzymatic degradation of a starch-based ternary semi-interpenetrating network for gastrointestinal drug delivery. J Appl Polym Sci 100:2975–2984. https://doi.org/10.1002/app.22897
Bajpai SK, Saxena S, Dubey S (2006) The flow-through diffusion cell (FTDC) method: a novel approach to in vitro drug release studies. Polym Int 55:12–18. https://doi.org/10.1002/pi.1954
Baker JP, Blanch HW, Prausnitz JM (1994) Equilibrium swelling properties of weakly lonizable 2-hydroxyethyl methacrylate (HEMA)-based hydrogels. J Appl Polym Sci 52:783–788
Bakrudeen HB, Sudarvizhi C, Reddy BSR (2016) Starch nanocrystals based hydrogel: construction, characterizations and transdermal application. Mater Sci Eng C 68:880–889. https://doi.org/10.1016/j.msec.2016.07.018
Bardajee GR, Hooshyar Z (2013) A novel biocompatible magnetic iron oxide nanoparticles/hydrogel based on poly (acrylic acid) grafted onto starch for controlled drug release. J Polym Res 20:298. https://doi.org/10.1007/s10965-013-0298-y
Bernardos A et al (2010) Enzyme-responsive intracellular controlled release using nanometric silica mesoporous supports capped with "saccharides". ACS Nano 4:6353–6368. https://doi.org/10.1021/nn101499d
Bie P, Chen L, Li X, Li L (2016) Characterization of concanavalin A-conjugated resistant starch acetate bioadhesive film for oral colon-targeting microcapsule delivery system. Ind Crops Prod 84:320–329. https://doi.org/10.1016/j.indcrop.2016.02.023
Björses K, Faxälv L, Montan C, Wildt-Persson K, Fyhr P, Holst J, Lindahl TL (2011) vitro and in vivo evaluation of chemically modified degradable starch microspheres for topical haemostasis. Acta Biomater 7:2558–2565. https://doi.org/10.1016/j.actbio.2011.03.003
Bolourchian N, Hadidi N, Foroutan SM, Shafaghi B (2009) Development and optimization of a sublingual tablet formulation for physostigmine salicylate. Acta Pharmaceutica 59:301–312. https://doi.org/10.2478/v10007-009-0028-5

Callens C, Remon JP (2000) Evaluation of starch-maltodextrin-Carbopol(®) 974 P mixtures for the nasal delivery of insulin in rabbits. J Control Release 66:215–220. https://doi.org/10.1016/s0168-3659(99)00271-0

Carbinatto FM, Ribeiro TS, Colnago LA, Evangelista RC, Cury BS (2016) Preparation and characterization of amylose inclusion complexes for drug delivery applications. J Pharm Sci 105:231–241

Carter R, Cooke TG, Hemingway D, McArdle CS, Angerson W (1992) The combination of degradable starch microspheres and angiotensin II in the manipulation of drug delivery in an animal model of colorectal metastasis. Br J Cancer 65:37–39

Chen D, Song P, Jiang F, Meng X, Sui W, Shu C, Wan L-J (2013) pH-responsive mechanism of a deoxycholic acid and folate comodified chitosan micelle under cancerous environment. J Phys Chem B 117:1261–1268

Chen M, Gao C, Lü S, Chen Y, Liu M (2016a) Dual redox-triggered shell-sheddable micelles self-assembled from mPEGylated starch conjugates for rapid drug release RSC. Advances 6:9164–9174

Chen M, Gao C, Lü S, Chen Y, Liu M (2016b) Preparation of redox-sensitive, core-crosslinked micelles self-assembled from mPEGylated starch conjugates: remarkable extracellular stability and rapid intracellular drug release. RSC Adv 6:46159–46169

Clara I, Natchimuthu N (2017) Hydrogels based on starch-g-poly(sodium-2-acrylamido-2-methyl-1-propane sulfonate-co-methacrylic acid) as controlled drug delivery systems. Starch - Stärke 69:1600177-n/a. https://doi.org/10.1002/star.201600177

Codd JE, Deasy PB (1998) Formulation development and in vivo evaluation of a novel bioadhesive lozenge containing a synergistic combination of antifungal agents. Int J Pharm 173:13–24. https://doi.org/10.1016/s0378-5173(98)00228-2

Constantin M, Fundueanu G, Cortesi R, Esposito E, Nastruzzi C (2003) Aminated polysaccharide microspheres as DNA delivery systems. Drug Delivery 10:139–149. https://doi.org/10.1080/10717540390215537

Cortesi R, Nastruzzi C, Davis S (1998) Sugar cross-linked gelatin for controlled release: microspheres and disks. Biomaterials 19:1641–1649

Dakhil S, Ensminger W, Cho K, Niederhuber J, Doan K, Wheeler R (1982) Improved regional selectivity of hepatic arterial BCNU with degradable microspheres. Cancer 50:631–635

Das NG, Das SK (2004) Development of mucoadhesive dosage forms of buprenorphine for sublingual drug delivery. Drug Delivery: J Delivery Target Ther Agents 11:89–95. https://doi.org/10.1080/10717540490280688

Desai KGH (2005) Preparation and characteristics of high-amylose corn starch/pectin blend microparticles: a technical note. AAPS PharmSciTech 6:E202–E208. https://doi.org/10.1208/pt060230

Desai KG (2007) Properties of tableted high-amylose corn starch-pectin blend microparticles intended for controlled delivery of diclofenac sodium. J Biomater Appl 21:217–233. https://doi.org/10.1177/0885328206056771

Devy J, Balasse E, Kaplan H, Madoulet C, Andry MC (2006) Hydroxyethylstarch microcapsules: a preliminary study for tumor immunotherapy application. Int J Pharm 307:194–200. https://doi.org/10.1016/j.ijpharm.2005.09.035

Dong D et al (2016) In Situ "Clickable" Zwitterionic Starch-based hydrogel for 3D cell encapsulation. ACS Appl Mater Interfaces 8:4442–4455

Dragan ES, Apopei DF (2013) Multiresponsive macroporous semi-IPN composite hydrogels based on native or anionically modified potato starch. Carbohydr Polym 92:23–32. https://doi.org/10.1016/j.carbpol.2012.08.082

Dragan ES, Perju MM, Dinu MV (2012) Preparation and characterization of IPN composite hydrogels based on polyacrylamide and chitosan and their interaction with ionic dyes. Carbohydr Polym 88:270–281

Dragan ES, Apopei Loghin DF, Cocarta A-I, Doroftei M (2016) Multi-stimuli-responsive semi-IPN cryogels with native and anionic potato starch entrapped in poly(*N*,*N*-dimethylaminoethyl methacrylate) matrix and their potential in drug delivery. React Funct Polym 105:66–77. https://doi.org/10.1016/j.reactfunctpolym.2016.05.015

Eid M (2008) In vitro release studies of vitamin B_{12} from poly *N*-vinyl pyrrolidone/starch hydrogels grafted with acrylic acid synthesized by gamma radiation. Nucl Instrum Methods Phys Res Section B: Beam Interact Mater Atoms 266:5020–5026. https://doi.org/10.1016/j.nimb.2008.09.003

El-Arnaouty M, Eid M, Abdel Ghaffar A (2015a) Radiation synthesis of stimuli responsive microporous hydrogels for controlled drug release of aspirin. Polym-Plast Technol Eng 54:1215–1222

El-Arnaouty MB, Abdel Ghaffar AM, Aboulfotouh ME, Taher NH, Taha AA (2015b) Radiation synthesis and characterization of Poly(butyl methacrylate/acrylamide) copolymeric hydrogels and heparin controlled drug release. Polym Bull 72:2739–2756. https://doi.org/10.1007/s00289-015-1433-1

Elvira C, Mano JF, San Roman J, Reis RL (2002) Starch-based biodegradable hydrogels with potential biomedical applications as drug delivery systems. Biomaterials 23:1955–1966. https://doi.org/10.1016/s0142-9612(01)00322-2

Engelberth SA, Hempel N, Bergkvist M (2015) Chemically modified dendritic starch: a novel nanomaterial for siRNA delivery. Bioconjug Chem 26:1766–1774

Ensminger WD, Gyves JW, Stetson P, Walker-Andrews S (1985) Phase I study of hepatic arterial degradable starch microspheres and mitomycin. Cancer Res 45:4464–4467

Fallingborg J, Christensen LA, Jacobsen BA, Rasmussen SN (1993) Very low intraluminal colonic pH in patients with active ulcerative colitis. Dig Dis Sci 38:1989–1993

Faroongsarng D, Sukonrat P (2008) Thermal behavior of water in the selected starch- and cellulose-based polymeric hydrogels. Int J Pharm 352:152–158. https://doi.org/10.1016/j.ijpharm.2007.10.022

Fathi M, Entezami AA, Arami S, Rashidi M-R (2015) Preparation of N-Isopropylacrylamide/Itaconic acid magnetic nanohydrogels by modified starch as a crosslinker for anticancer drug carriers. Int J Polym Mater Polym Biomater 64:541–549. https://doi.org/10.1080/00914037.2014.996703

Freire C, Podczeck F, Ferreira D, Veiga F, Sousa J, Pena A (2010) Assessment of the in-vivo drug release from pellets film-coated with a dispersion of high amylose starch and ethylcellulose for potential colon delivery. J Pharm Pharmacol 62:55–61. https://doi.org/10.1211/jpp.62.01.0005

Fundueanu G, Constantin M, Ascenzi P, Simionescu BC (2010) An intelligent multicompartmental system based on thermo-sensitive starch microspheres for temperature-controlled release of drugs. Biomed Microdevices 12:693–704. https://doi.org/10.1007/s10544-010-9422-5

Funke U, Lindhauer MG (2001) Effect of reaction conditions and alkyl chain lengths on the properties of hydroxyalkyl starch ethers. Starch - Stärke 53:547–554

Galgatte UC, Khanchandani SS, Jadhav YG, Chaudhari PD (2013) Investigation of different polymers, plasticizers and superdisintegrating agents alone and in combination for use in the formulation of fast dissolving oral films. Int J PharmTech Res 5:1465–1472

Galloway G, Biliaderis C, Stanley D (1989) Properties and structure of amylose-glyceryl monostearate complexes formed in solution or on extrusion of wheat flour. J Food Sci 54:950–957

Ganta S, Devalapally H, Shahiwala A, Amiji M (2008) A review of stimuli-responsive nanocarriers for drug and gene delivery. J Controlled Release 126:187–204

Gils PS, Ray D, Mohanta GP, Manavalan R, Sahoo PK (2009) Designing of new acrylic based macroporous superabsorbent polymer hydrogel and its suitability for drug delivery. Int J Pharm Pharm Sci 1:43–54

Godbey W, Wu KK, Mikos AG (1999) Tracking the intracellular path of poly (ethylenimine)/DNA complexes for gene delivery. Proc Natl Acad Sci 96:5177–5181

Godet M, Bouchet B, Colonna P, Gallant D, Buleon A (1996) Crystalline amylose-fatty acid complexes: morphology and crystal thickness. J Food Sci 61:1196–1201

Grinberg O, Gedanken A (2010) The development and characterization of starch microspheres prepared by a sonochemical method for the potential drug delivery of insulin. Macromol Chem Phys 211:924–931. https://doi.org/10.1002/macp.200900613

Guilherme MR et al (2012) Albumin release from a brain-resembling superabsorbent magnetic hydrogel based on starch. Soft Matter 8:6629–6637. https://doi.org/10.1039/c2sm25638j

Güler MA, Gök MK, Figen AK, Özgümüş S (2015) Swelling, mechanical and mucoadhesion properties of Mt/starch-g-PMAA nanocomposite hydrogels. Appl Clay Sci 112:44–52

Hamdi G, Ponchel G, Duchêne D (1998) An original method for studying in vitro the enzymatic degradation of cross-linked starch microspheres. J Controlled Release 55:193–201. https://doi.org/10.1016/S0168-3659(98)00055-8

Hamidian H, Tavakoli T (2016) Preparation of a new Fe_3O_4/starch-g-polyester nanocomposite hydrogel and a study on swelling and drug delivery properties. Carbohydr Polym 144:140–148. https://doi.org/10.1016/j.carbpol.2016.02.048

Han X, Zhang X, Yin Q, Hu J, Liu H, Hu Y (2013) Thermoresponsive diblock copolymer with tunable soluble–insoluble and soluble–insoluble–soluble transitions. Macromol Rapid Commun 34:574–580

Hata H, Matsuzaki H, Sanada I, Takatsuk K (1990) Genetic analysis of amylase-producing cell lines: ectopic activation of the amylase gene by translocation. Jpn J Clin Oncol 20:246–251

Haug A, Larsen B (1963) The solubility of alginate at low pH. Acta Chem Scand 17:1653–1662

Heinemann C, Zinsli M, Renggli A, Escher F, Conde-Petit B (2005) Influence of amylose-flavor complexation on build-up and breakdown of starch structures in aqueous food model systems LWT-Food. Sci Technol 38:885–894

Helaly FM, Khalaf AI, El Nashar D (2013) Starch cellulose acetate co-acrylate (SCAA) polymer as a drug carrier. Res Chem Intermed 39:3209–3220. https://doi.org/10.1007/s11164-012-0833-1

Holmberg K, Bjork E, Bake B, Edman P (1994) Influence of degradable starch microspheres on the human nasal mucosa. Rhinology 32:74–77

Hong Y, Liu G, Gu Z (2016) Recent advances of starch-based excipients used in extended-release tablets: a review. Drug Delivery 23:12–20

Huang Y, Ding S, Liu M, Gao C, Yang J, Zhang X, Ding B (2013) Ultra-small and anionic starch nanospheres: Formation and vitro thrombolytic behavior study. Carbohydr Polym 96:426–434. https://doi.org/10.1016/j.carbpol.2013.04.013

Huang Y, Liu M, Gao C, Yang J, Zhang X, Zhang X, Liu Z (2013) Ultra-small and innocuous cationic starch nanospheres: preparation, characterization and drug delivery study. Int J Biol Macromol 58:231–239. https://doi.org/10.1016/j.ijbiomac.2013.04.006

Hui A, Sheardown H, Jones L (2012) Acetic and acrylic acid molecular imprinted model silicone hydrogel materials for ciprofloxacin-HCl delivery. Materials 5:85–107

Huo W, Xie G, Zhang W, Wang W, Shan J, Liu H, Zhou X (2016) Preparation of a novel chitosan-microcapsules/starch blend film and the study of its drug-release mechanism. Int J Biol Macromol 87:114–122. https://doi.org/10.1016/j.ijbiomac.2016.02.049

Ismail H, Irani M, Ahmad Z (2013) Starch-based hydrogels: present status and applications. Int J Polym Mater Polym Biomater 62:411–420. https://doi.org/10.1080/00914037.2012.719141

Jane J-l, Robyt JF (1984) Structure studies of amylose-V complexes and retro-graded amylose by action of alpha amylases, and a new method for preparing amylodextrins. Carbohydr Res 132:105–118

Jie Z, Jian Z, Allan ED, Victor CY (2013) Magnetic tumor targeting of β-glucosidase immobilized iron oxide nanoparticles. Nanotechnology 24:375102

Jin S, Liu M, Zhang F, Chen S, Niu A (2006) Synthesis and characterization of pH-sensitivity semi-IPN hydrogel based on hydrogen bond between poly(N-vinylpyrrolidone) and poly(acrylic acid). Polymer 47:1526–1532. http://dx.doi.org/10.1016/j.polymer.2006.01.009

Joseph J, Viney S, Beck B, Strange C, Sahn SA, Basran GS (1992) A prospective study of amylase-rich pleural effusions with special reference to amylase isoenzyme analysis. Chest 102:1455–1459. https://doi.org/10.1378/chest.102.5.1455

Jovic M, Sharma M, Rahajeng J, Caplan S (2010) The early endosome: a busy sorting station for proteins at the crossroads. Histol Histopathol 25:99–112

Ju B, Yan D, Zhang S (2012) Micelles self-assembled from thermoresponsive 2-hydroxy-3-butoxypropyl starches for drug delivery. Carbohydr Polym 87:1404–1409. https://doi.org/10.1016/j.carbpol.2011.09.028

Ju B, Zhang C, Zhang S (2014) Thermoresponsive starch derivates with widely tuned LCSTs by introducing short oligo(ethylene glycol) spacers. Carbohydr Polym 108:307–312. https://doi.org/10.1016/j.carbpol.2014.02.057

Karrout Y et al (2011) Peas starch-based film coatings for site-specific drug delivery to the colon. J Appl Polym Sci 119:1176–1184. https://doi.org/10.1002/app.32802

Kichler A (2004) Gene transfer with modified polyethylenimines. J Gene Med 6:S3–S10

Kim J-Y, Huber KC (2016) Preparation and characterization of corn starch-β-carotene composites. Carbohydr Polym 136:394–401

Kim H-Y, Park SS, Lim S-T (2015) Preparation, characterization and utilization of starch nanoparticles Colloids Surf B: Biointerfaces 126:607–620. https://doi.org/10.1016/j.colsurfb.2014.11.011

Kommareddy S, Amiji M (2005) Preparation and evaluation of thiol-modified gelatin nanoparticles for intracellular DNA delivery in response to glutathione. Bioconjug Chem 16:1423–1432

Kovacs AF, Turowski B (2002) Chemoembolization of oral and oropharyngeal cancer using a high-dose cisplatin crystal suspension and degradable starch microspheres. Oral Oncol 38:87–95. https://doi.org/10.1016/s1368-8375(01)00088-4

Krishnaiah Y, Reddy PB, Satyanarayana V, Karthikeyan R (2002) Studies on the development of oral colon targeted drug delivery systems for metronidazole in the treatment of amoebiasis. Int J Pharm 236:43–55

Kumar P, Ganure AL, Subudhi BB, Shukla S (2015) Preparation and characterization of pH-sensitive methyl methacrylate-g-starch/hydroxypropylated starch hydrogels: in vitro and in vivo study on release of esomeprazole magnesium. Drug Delivery Transl Res 5:243–256

Kwag DS, Oh KT, Lee ES (2014) Facile synthesis of multilayered polysaccharidic vesicles. J Controlled Release 187:83–90. https://doi.org/10.1016/j.jconrel.2014.05.032

Larionov NV, Ponchel G, Duchene D, Larionova NI (1999) Biodegradable cross-linked starch/protein microcapsules containing proteinase inhibitor for oral protein administration. Int J Pharm 189:171–178. https://doi.org/10.1016/s0378-5173(99)00249-5

Le Corre D, Bras J, Dufresne A (2010) Starch nanoparticles: a review. Biomacromol 11:1139–1153. https://doi.org/10.1021/bm901428y

Lechardeur D, Lukacs GL (2002) Intracellular barriers to non-viral gene transfer. Curr Gene Ther 2:183–194

Lechardeur D et al (1999) Metabolic instability of plasmid DNA in the cytosol: a potential barrier to gene transfer. Gene Ther 6:482–497

Lesmes U, Barchechath J, Shimoni E (2008) Continuous dual feed homogenization for the production of starch inclusion complexes for controlled release of nutrients. Innov Food Sci Emerg Technol 9:507–515. https://doi.org/10.1016/j.ifset.2007.12.008

Li Y, de Vries R, Slaghek T, Timmermans J, Cohen Stuart MA, Norde W (2009) Preparation and characterization of oxidized starch polymer microgels for encapsulation and controlled release of functional ingredients. Biomacromol 10:1931–1938. https://doi.org/10.1021/bm900337n

Li Y, Kleijn JM, Stuart MAC, Slaghek T, Timmermans J, Norde W (2011a) Mobility of lysozyme inside oxidized starch polymer microgels. Soft Matter 7:1926–1935. https://doi.org/10.1039/c0sm00962h

Li Y, Zhang Z, Van Leeuwen HP, Cohen Stuart MA, Norde W, Kleijn JM (2011b) Uptake and release kinetics of lysozyme in and from an oxidized starch polymer microgel. Soft Matter 7:10377–10385. https://doi.org/10.1039/c1sm06072d

Li Y, Liu C, Tan Y, Xu K, Lu C, Wang P (2014) situ hydrogel constructed by starch-based nanoparticles via a Schiff base reaction. Carbohydr Polym 110:87–94. https://doi.org/10.1016/j.carbpol.2014.03.058

Li K et al (2015) Mulberry-like dual-drug complicated nanocarriers assembled with apogossypolone amphiphilic starch micelles and doxorubicin hyaluronic acid nanoparticles for tumor combination

and targeted therapy. Biomaterials 39:131–144. https://doi.org/10.1016/j.biomaterials.2014.10.073
Li J, Shin GH, Lee IW, Chen X, Park HJ (2016) Soluble starch formulated nanocomposite increases water solubility and stability of curcumin. Food Hydrocolloids 56:41–49
Likhitkar S, Bajpai AK (2012) Magnetically controlled release of cisplatin from superparamagnetic starch nanoparticles. Carbohydr Polym 87:300–308. https://doi.org/10.1016/j.carbpol.2011.07.053
Likhitkar S, Bajpai AK (2014) An in vitro experimental approach to study magnetically targeted release of methotrexate from superparamagnetic starch nanocarriers. Int J Polym Mater and Polym Biomater 63:941–950. https://doi.org/10.1080/00914037.2014.886232
Lima-Tenório MK, Tenório-Neto ET, Guilherme MR, Garcia FP, Nakamura CV, Pineda EAG, Rubira AF (2015) Water transport properties through starch-based hydrogel nanocomposites responding to both pH and a remote magnetic field. Chem Eng J 259:620–629. https://doi.org/10.1016/j.cej.2014.08.045
Liu L, Fishman ML, Kost J, Hicks KB (2003) Pectin-based systems for colon-specific drug delivery via oral route. Biomaterials 24:3333–3343
Liu C, Gan X, Chen Y (2011) A novel pH-sensitive hydrogels for potential colon-specific drug delivery: characterization and in vitro release studies. Starch - Stärke 63:503–511. https://doi.org/10.1002/star.201000120
Liu K, Wang Y, Li H, Duan Y (2015) A facile one-pot synthesis of starch functionalized graphene as nano-carrier for pH sensitive and starch-mediated drug delivery. Colloids Surf B: Biointerfaces 128:86–93. https://doi.org/10.1016/j.colsurfb.2015.02.010
Liu M et al (2016) Fabrication and biological imaging application of AIE-active luminescent starch based nanoprobes. Carbohydr Polym 142:38–44
Lorelius LE, Benedetto AR, Blumhardt R, Gaskill HV 3rd, Lancaster JL, Stridbeck H (1984) Enhanced drug retention in VX2 tumors by use of degradable starch microspheres. Invest Radiol 19:212–215
Mahkam M (2010) Modified chitosan cross-linked starch polymers for oral insulin delivery. J Bioact Compatible Polym 25:406–418. https://doi.org/10.1177/0883911510369038
Mamada A, Tanaka T, Kungwatchakun D, Irie M (1990) Photoinduced phase transition of gels. Macromolecules 23:1517–1519
Mehyar GF, Liu Z, Han JH (2008) Dynamics of antimicrobial hydrogels in physiological saline. Carbohydr Polym 74:92–98. https://doi.org/10.1016/j.carbpol.2008.01.023
Meng F, Hennink WE, Zhong Z (2009) Reduction-sensitive polymers and bioconjugates for biomedical applications. Biomaterials 30:2180–2198
Misic Z, Urbani R, Pfohl T, Muffler K, Sydow G, Kuentz M (2014) Understanding biorelevant drug release from a novel thermoplastic capsule by considering microstructural formulation changes during hydration. Pharm Res 31:194–203. https://doi.org/10.1007/s11095-013-1152-y
Mohanty DP, Biswal S, Nayak L (2015) Preparation of starch-chitosan nanocomposites for control drug release of curcumin. Int J Curr Eng Technol 5:336–343
Nitta SK, Numata K (2013) Biopolymer-based nanoparticles for drug/gene delivery and tissue engineering. Int J Mol Sci 14:1629–1654. https://doi.org/10.3390/ijms14011629
Noakes M, Clifton PM, Nestel PJ, Le Leu R, McIntosh G (1996) Effect of high-amylose starch and oat bran on metabolic variables and bowel function in subjects with hypertriglyceridemia. Am J Clin Nutr 64:944–951
Noga M et al (2013) The effect of molar mass and degree of hydroxyethylation on the controlled shielding and deshielding of hydroxyethyl starch-coated polyplexes. Biomaterials 34:2530–2538. https://doi.org/10.1016/j.biomaterials.2012.12.025
Noga M, Edinger D, Wagner E, Winter G, Besheer A (2014) Characterization and compatibility of hydroxyethyl starch-polyethylenimine copolymers for DNA delivery. J Biomater Sci-Polym Edn 25:855–871. https://doi.org/10.1080/09205063.2014.910152

Patten GS, Augustin MA, Sanguansri L, Head RJ, Abeywardena MY (2009) Site specific delivery of microencapsulated fish oil to the gastrointestinal tract of the rat. Dig Dis Sci 54:511–521. https://doi.org/10.1007/s10620-008-0379-7

Peppas NA (1997) Hydrogels and drug delivery. Curr Opin Colloid Interface Sci 2:531–537

Pereira AGB, Fajardo AR, Nocchi S, Nakamura CV, Rubira AF, Muniz EC (2013) Starch-based microspheres for sustained-release of curcumin: preparation and cytotoxic effect on tumor cells. Carbohydr Polym 98:711–720. https://doi.org/10.1016/j.carbpol.2013.06.013

Plamper FA, Ruppel M, Schmalz A, Borisov O, Ballauff M, Müller AH (2007) Tuning the thermoresponsive properties of weak polyelectrolytes: aqueous solutions of star-shaped and linear poly (*N*,*N*-dimethylaminoethyl methacrylate) Macromolecules 40:8361–8366

Pourjamal K, Fathi M, Entezami AA, Hasanzadeh M, Shadjou N (2016) Superabsorbent nanohydrogels of poly (N-isopropyl acrylamide-co-itaconic acid) grafted on starch: Synthesis and swelling study. Nano LIFE 06:1650005

Pourjavadi A, Ebrahimi AA, Barzegar S (2013) Preparation and evaluation of bioactive and compatible starch based superabsorbent for oral drug delivery systems. J Drug Deliv Sci Technol 23:511–517

Pourjavadi A, Tehrani ZM, Hosseini SH (2015) Dendritic magnetite decorated by pH-responsive PEGylated starch: a smart multifunctional nanocarrier for the triggered release of anti-cancer drugs. RSC Advances 5:48586–48595

Putseys J, Lamberts L, Delcour J (2010) Amylose-inclusion complexes: formation, identity and physico-chemical properties. J Cereal Sci 51:238–247

Ramasubbu N, Paloth V, Luo Y, Brayer GD, Levine MJ (1996) Structure of human salivary α-amylase at 1.6 A resolution: implications for its role in the oral cavity. Acta Crystallogr Sect D: Struct Biol 52:435–446. https://doi.org/10.1107/S0907444995014119

Rodrigues A, Emeje M (2012) Recent applications of starch derivatives in nanodrug delivery. Carbohydr Polym 87:987–994. https://doi.org/10.1016/j.carbpol.2011.09.044

Saboktakin MR, Maharramov A, Ramazanov MA, Mahkam M (2007) Modification of carboxymethyl starch as nano carriers for oral drug delivery. Nat Sci 5:30–36

Saboktakin MR, Maharramov A, Ramazanov MA (2009a) pH-sensitive starch hydrogels via free radical graft copolymerization, synthesis and properties. Carbohydr Polym 77:634–638. https://doi.org/10.1016/j.carbpol.2009.02.004

Saboktakin MR, Maharramov A, Ramazanov MA (2009b) Synthesis and characterization of superparamagnetic nanoparticles coated with carboxymethyl starch (CMS) for magnetic resonance imaging technique. Carbohydr Polym 78:292–295. https://doi.org/10.1016/j.carbpol.2009.03.042

Saboktakin MR, Tabatabaie RM, Maharramov A, Ramazanov MA (2011) Synthesis and in vitro evaluation of carboxymethyl starch-chitosan nanoparticles as drug delivery system to the colon. Int J Biol Macromol 48:381–385. https://doi.org/10.1016/j.ijbiomac.2010.10.005

Sadeghi M, Hosseinzadeh H (2008) Synthesis of starch-poly(sodium acrylate-co-acrylamide) superabsorbent hydrogel with salt and pH-responsiveness properties as a drug delivery system. J Bioact Compatible Polym 23:381–404. https://doi.org/10.1177/0883911508093504

Saikia C, Hussain A, Ramteke A, Sharma HK, Maji TK (2014) Crosslinked thiolated starch coated Fe_3O_4 magnetic nanoparticles: effect of montmorillonite and crosslinking density on drug delivery properties. Starch - Stärke 66:760–771

Saikia C, Hussain A, Ramteke A, Sharma HK, Maji TK (2015) Carboxymethyl starch-chitosan-coated iron oxide magnetic nanoparticles for controlled delivery of isoniazid. J Microencapsul 32:29–39

Saikia C, Das MK, Ramteke A, Maji TK (2016) Effect of crosslinker on drug delivery properties of curcumin loaded starch coated iron oxide nanoparticles. Int J Biol Macromol 93:1121–1132. https://doi.org/10.1016/j.ijbiomac.2016.09.043

Saikia C, Das MK, Ramteke A, Maji TK (2017) Evaluation of folic acid tagged aminated starch/ZnO coated iron oxide nanoparticles as targeted curcumin delivery system. Carbohydr Polym 157:391–399. https://doi.org/10.1016/j.carbpol.2016.09.087

Sarko A, Wu HC (1978) The crystal structures of A-, B- and C-Polymorphs of amylose and starch. Starch - Stärke 30:73–78

Sasaki Y, Hada R, Nakajima H, Fukuda S, Munakata A (1997) Improved localizing method of radiopill in measurement of entire gastrointestinal pH profiles: colonic luminal pH in normal subjects and patients with Crohn's disease. Am J Gastroenterol 92

Schmitt H, Creton N, Prashantha K, Soulestin J, Lacrampe MF, Krawczak P (2015) Melt-blended halloysite nanotubes/wheat starch nanocomposites as drug delivery system. Polym Eng Sci 55:573–580. https://doi.org/10.1002/pen.23919

Schwoerer A, Harling S, Menzel H, Daniels R (2008) Release behaviour of hydrogel microparticles based on hydroxyl-ethyl-starch as a drug delivery system for proteins. J. Controlled Release. 132:e16–e17. https://doi.org/10.1016/j.jconrel.2008.09.050

Schwoerer ADA, Harling S, Scheibe K, Menzel H, Daniels R (2009) Influence of degree of substitution of HES-HEMA on the release of incorporated drug models from corresponding hydrogels. Eur. J. Pharm. Biopharm. 73:351–356. https://doi.org/10.1016/j.ejpb.2009.08.003

Setty CM, Deshmukh AS, Badiger AM (2014) Hydrolyzed polyacrylamide grafted maize starch based microbeads: application in pH responsive drug delivery. Int J Biol Macromol 70:1–9. https://doi.org/10.1016/j.ijbiomac.2014.06.027

Shaikh MMM, Gramopadhye NS, Wardole AA, Lonikar SV (2015) Starch-acrylic acid hydrogel: synthesis, characterization and drug release study

Shalviri A et al (2012) pH-Dependent doxorubicin release from terpolymer of starch, polymethacrylic acid and polysorbate 80 nanoparticles for overcoming multi-drug resistance in human breast cancer cells. Eur J Pharm Biopharm 82:587–597. https://doi.org/10.1016/j.ejpb.2012.09.001

Shalviri A, Chan HK, Raval G, Abdekhodaie MJ, Liu Q, Heerklotz H, Wu XY (2013) Design of pH-responsive nanoparticles of terpolymer of poly(methacrylic acid), polysorbate 80 and starch for delivery of doxorubicin. Colloids Surf B: Biointerfaces 101:405–413. https://doi.org/10.1016/j.colsurfb.2012.07.015

Shenoy DB, Amiji MM (2005) Poly (ethylene oxide)-modified poly (ε-caprolactone) nanoparticles for targeted delivery of tamoxifen in breast cancer. Int J Pharm 293:261–270

Shenoy D, Little S, Langer R, Amiji M (2005) Poly (ethylene oxide)-modified poly (β-amino ester) nanoparticles as a pH-sensitive system for tumor-targeted delivery of hydrophobic drugs: part 2. In vivo distribution and tumor localization studies. Pharm Res 22:2107–2114

Shi J, Votruba AR, Farokhzad OC, Langer R (2010) Nanotechnology in drug delivery and tissue engineering: from discovery to applications. Nano Lett 10:3223–3230. https://doi.org/10.1021/nl102184c

Shiba H, Okamoto T, Futagawa Y, Misawa T, Yanaga K, Ohashi T, Eto Y (2006) Adenovirus vector-mediated gene transfer using degradable starch microspheres for hepatocellular carcinoma in rats. J Surg Res 133:193–196. https://doi.org/10.1016/j.jss.2005.10.023

Shiga T, Hirose Y, Okada A, Kurauchi T (1992) Electric field-associated deformation of polyelectrolyte gel near a phase transition point. J Agric Food Chem 46:635–640

Sieradzki R, Traitel T, Goldbart R, Geresh S, Kost J (2014) Tailoring quaternized starch as a non-viral carrier for gene delivery applications. Polym Adv Technol 25:552–561. https://doi.org/10.1002/pat.3277

Sigurdson ER, Ridge JA, Daly JM (1986) Intra-arterial infusion of doxorubicin with degradable starch microspheres: improvement of hepatic tumor drug uptake. Arch Surg 121:1277–1281. https://doi.org/10.1001/archsurg.1986.01400110067011

Sinha V, Kumria R (2001) Polysaccharides in colon-specific drug delivery. Int J Pharm 224:19–38

Situ W, Chen L, Wang X, Li X (2014) Resistant starch film-coated microparticles for an oral colon-specific polypeptide delivery system and its release behaviors. J Agric Food Chem 62:3599–3609. https://doi.org/10.1021/jf500472b

Situ W, Li X, Liu J, Chen L (2015) Preparation and characterization of glycoprotein-resistant starch complex as a coating material for oral bioadhesive microparticles for colon-targeted polypeptide delivery. J Agric Food Chem 63:4138–4147

Smith BW, Roe JH (1949) A photometric method for the determination of α-amylase in blood and urine, with use of the starch-iodine color. J Biol Chem 179:53–59

Soares GA, Carbinatto FM, Cury BSF, Evangelista RC (2013a) Effect of drying technique on some physical properties of cross-linked high amylose/pectin mixtures. Drug Dev Ind Pharm 39:284–289. https://doi.org/10.3109/03639045.2012.679278

Soares GA, Castro ADD, Cury BSF, Evangelista RC (2013b) Blends of cross-linked high amylose starch/pectin loaded with diclofenac. Carbohydr Polym 91:135–142. https://doi.org/10.1016/j.carbpol.2012.08.014

Subramanian SB, Francis AP, Devasena T (2014) Chitosan-starch nanocomposite particles as a drug carrier for the delivery of bis-desmethoxy curcumin analog. Carbohydr Polym 114:170–178. https://doi.org/10.1016/j.carbpol.2014.07.053

Surini S, Anggriani V, Anwar E (2009) Study of mucoadhesive microspheres based on pregelatinized cassava starch succinate as a new carrier for drug delivery. J Med Sci 9:249–256. https://doi.org/10.3923/jms.2009.249.256

Tan HL, Wong YY, Muniyandy S, Hashim K, Pushpamalar J (2016) Carboxymethyl sago pulp/carboxymethyl sago starch hydrogel: effect of polymer mixing ratio and study of controlled drug release. J Appl Polym Sci 133:43652. https://doi.org/10.1002/app.43652

Tan Y, Xu K, Li Y, Sun S, Wang P (2010) A robust route to fabricate starch esters vesicles. Chem Commun 46:4523–4525. https://doi.org/10.1039/C000471E

Thakur BR, Singh RK, Handa AK, Rao M (1997) Chemistry and uses of pectin—a review. Crit Rev Food Sci Nutr 37:47–73

Thiele C, Loretz B, Lehr C-M (2017) Biodegradable starch derivatives with tunable charge density—synthesis, characterization, and transfection efficiency. Drug Delivery Transl Res 7:252–258. https://doi.org/10.1007/s13346-016-0333-8

Thom AK, Reilly CA, Deveney CW, Hansell JR, Neufeld GR, Daly JM (1988) The use of quantitative perfusion fluorometry to measure relative tumor and liver blood flow after transient microembolization. J Surg Res 45:128–133

Thombre NA, Chaudhari MR, Kadam SS (2009) Preparation and characterization of rofecoxib microspheres using cross-linked starch as novel drug delivery system. Int J PharmTech Res 1:1394–1402

Thompson DB (2000) Strategies for the manufacture of resistant starch. Trends Food Sci Technol 11:245–253

Torres FG, Commeaux S, Troncoso OP (2013) Starch-based biomaterials for wound-dressing applications. Starch - Stärke 65:543–551. https://doi.org/10.1002/star.201200259

Tuma RF, Forsberg JO, Agerup B (1982) Enhanced uptake of actinomycin D in the dog kidney by simultaneous injection of degradable starch microspheres into the renal artery. Cancer 50:1–5

Tuovinen L, Peltonen S, Liikola M, Hotakainen M, Lahtela-Kakkonen M, Poso A, Jarvinen K (2004a) Drug release from starch-acetate microparticles and films with and without incorporated alpha-amylase. Biomaterials 25:4355–4362. https://doi.org/10.1016/j.biomaterials.2003.11.026

Tuovinen L et al (2004b) Starch acetate microparticles for drug delivery into retinal pigment epithelium—in vitro study. J Controlled Release 98:407–413. https://doi.org/10.1016/j.jconrel.2004.05.016

Vakili MR, Rahneshin N (2013) Synthesis and characterization of novel stimuli-responsive hydrogels based on starch and L-aspartic acid. Carbohydra Polym 98:1624–1630. https://doi.org/10.1016/j.carbpol.2013.08.016

van Veen B, Pajander J, Zuurman K, Lappalainen R, Poso A, Frijlink HW, Ketolainen J (2005) The effect of powder blend and tablet structure on drug release mechanisms of hydrophobic starch acetate matrix tablets. Eur J Pharm Biopharm 61:149–157. https://doi.org/10.1016/j.ejpb.2005.04.007

Veiseh O, Tang BC, Whitehead KA, Anderson DG, Langer R (2015) Managing diabetes with nanomedicine: challenges and opportunities. Nat Rev Drug Discov 14:45–57

Wang Q, Hu X, Du Y, Kennedy JF (2010) Alginate/starch blend fibers and their properties for drug controlled release. Carbohydr Polym 82:842–847. https://doi.org/10.1016/j.carbpol.2010.06.004

Wang J, Liu H, Leng F, Zheng L, Yang J, Wang W, Huang CZ (2014) Autofluorescent and pH-responsive mesoporous silica for cancer-targeted and controlled drug release. Microporous Mesoporous Mater 186:187–193. https://doi.org/10.1016/j.micromeso.2013.11.006

Watts P, Smith A (2005) TARGIT technology: coated starch capsules for site-specific drug delivery into the lower gastrointestinal tract. Expert Opin Drug Delivery 2:159–167. https://doi.org/10.1517/17425247.2.1.159

Wolfert MA et al (1999) Polyelectrolyte vectors for gene delivery: influence of cationic polymer on biophysical properties of complexes formed with DNA. Bioconjug Chem 10:993–1004

Wu D-Q, Sun Y-X, Xu X-D, Cheng S-X, Zhang X-Z, Zhuo R-X (2008) Biodegradable and pH-sensitive hydrogels for cell encapsulation and controlled drug release. Biomacromol 9:1155–1162

Wu C, Yang J, Xu X, Gao C, Lü S, Liu M (2016) Redox-responsive core-cross-linked mPEGylated starch micelles as nanocarriers for intracellular anticancer drug release. Eur Polym J 83:230–243. https://doi.org/10.1016/j.eurpolymj.2016.08.018

Xiao H, Yang T, Lin Q, Liu GQ, Zhang L, Yu F, Chen Y (2016) Acetylated starch nanocrystals: preparation and antitumor drug delivery study. Int J Biol Macromol 89:456–464. https://doi.org/10.1016/j.ijbiomac.2016.04.037

Xu H, Cao W, Zhang X (2013) Selenium-containing polymers: promising biomaterials for controlled release and enzyme mimics. Acc Chem Res 46:1647–1658

Xuan Phuc N et al (2012) Iron oxide-based conjugates for cancer theragnostics. Adv Nat Sci Nanosci Nanotechnol 3:033001. https://doi.org/10.1088/2043-6262/3/3/033001

Yamada H, Loretz B, Lehr C-M (2014) Design of starch-graft-PEI polymers: an effective and biodegradable gene delivery platform. Biomacromol 15:1753–1761

Yang L, Chu JS, Fix JA (2002) Colon-specific drug delivery: new approaches and in vitro/in vivo evaluation. Int J Pharm 235:1–15

Yang Y, Jiang JS, Du B, Gan ZF, Qian M, Zhang P (2009) Preparation and properties of a novel drug delivery system with both magnetic and biomolecular targeting. J Mater Sci Mater Med 20:301–307. https://doi.org/10.1007/s10856-008-3577-0

Yang Y-H, Aloysius H, Inoyama D, Chen Y, Hu L-Q (2011) Enzyme-mediated hydrolytic activation of prodrugs. Acta Pharmaceutica Sinica B 1:143–159

Yang LQ, Zhang BF, Yi JZ, Liang JB, Liu YL, Zhang LM (2013) Preparation, characterization, and properties of amylose-ibuprofen inclusion complexes. Starch - Stärke 65:593–602. https://doi.org/10.1002/star.201200161

Yang J, Gao C, Lü S, Zhang X, Yu C, Liu M (2014a) Physicochemical characterization of amphiphilic nanoparticles based on the novel starch–deoxycholic acid conjugates and self-aggregates. Carbohydr Polym 102:838–845. https://doi.org/10.1016/j.carbpol.2013.10.081

Yang J, Huang Y, Gao C, Liu M, Zhang X (2014b) Fabrication and evaluation of the novel reduction-sensitive starch nanoparticles for controlled drug release. Colloids Surf B: Biointerfaces 115:368–376. https://doi.org/10.1016/j.colsurfb.2013.12.007

Yang JL, Gao CM, Lu SY, Wang XG, Chen MJ, Liu MZ (2014c) Novel self-assembled amphiphilic mPEGylated starch-deoxycholic acid polymeric micelles with pH-response for anticancer drug delivery. RSC Advances 4:55139–55149. https://doi.org/10.1039/c4ra07315k

Yoon H-S, Lee J, Lim S-T (2009) Utilization of retrograded waxy maize starch gels as tablet matrix for controlled release of theophylline. Carbohydr Polym 76:449–453

Yoshikawa T et al (1994) Antitumor effect of ischemia-reperfusion injury induced by transient embolization. Cancer Res 54:5033–5035

Zhang A et al (2013a) Redox-sensitive shell-crosslinked polypeptide-block-polysaccharide micelles for efficient intracellular anticancer drug delivery. Macromol Biosci 13:1249–1258

Zhang AP et al (2013b) Disulfide crosslinked PEGylated starch micelles as efficient intracellular drug delivery platforms. Soft Matter 9:2224–2233. https://doi.org/10.1039/c2sm27189c

Zhang LM, Yang C, Yan L (2005) Perspectives on: strategies to fabricate starch-based hydrogels with potential biomedical applications. J Bioact Compatible Polym 20:297–314. https://doi.org/10.1177/0883911505053882

Zhang L, Liu Y, Wu Z, Chen H (2009) Preparation and characterization of coacervate microcapsules for the delivery of antimicrobial oyster peptides. Drug Dev Ind Pharm 35:369–378. https://doi.org/10.1080/03639040802369255

Zhang Z, Shan HL, Chen L, He CL, Zhuang XL, Chen XS (2013c) Synthesis of pH-responsive starch nanoparticles grafted poly (L-glutamic acid) for insulin controlled release. Eur Polym J 49:2082–2091. https://doi.org/10.1016/j.eurpolymj.2013.04.032

Zhao Q, Sun J, Lin Y, Zhou Q (2010) Study of the properties of hydrolyzed polyacrylamide hydrogels with various pore structures and rapid pH-sensitivities. React Funct Polym 70:602–609

Zhao Q, Sun J, Wu X, Lin Y (2011) Macroporous double-network cryogels: formation mechanism, enhanced mechanical strength and temperature/pH dual sensitivity. Soft Matter 7:4284–4293

Zohreh N, Hosseini SH, Pourjavadi A (2016) Hydrazine-modified starch coated magnetic nanoparticles as an effective pH-responsive nanocarrier for doxorubicin delivery. J Ind Eng Chem 39:203–209. https://doi.org/10.1016/j.jiec.2016.05.029

Chapter 5
Starch-Based DDSs with Physiological Interactions

Abstract Drug delivery systems (DDSs) are designed to be able to precisely control the release rate and/or target drugs to specific body sites. However, the successful application of DDSs could be limited in the clinical application due to the complicated environment of the human body. This can be exampled by the short circulation time and low targeting efficiency of target-specific DDSs induced by the RES recognition, or low bioavailability of drugs caused by short residence time of transmucosal DDSs at the site of absorption. Regarding this, in this chapter, strategies to improve the performance and bioavailability of starch-based DDSs are discussed. Target-specific starch-based DDSs can be achieved by passive and active targeting. Another possible strategy is the physical targeting of drugs by external stimuli, such as magnetic field. Also, the role of starch and its derivatives in transmucosal DDSs to improve the bioavailability of drugs by interacting with the absorbing mucosa or prolonging the residence time of drugs in the absorbing tissues are highlighted.

Keywords Transmucosal starch-based drug delivery system · Bioadhesion · Widening tight junction effect · Target-specific starch-based drug delivery system · Blood hematocompatibility · Passive targeting · Active targeting · External targeting

5.1 Transmucosal Starch-Based DDSs

Regarding the application of transmucosal DDSs, the limitation of the available absorption area and the short residence time, in particular with the washing effect of saliva (Paderni et al. 2012) or the clearance by villus (Mailman et al. 1990), can hinder the absorption efficiency of drugs. Therefore, it would be advantageous to use means to enable an intimate contact of DDSs with the absorbing mucosa or prolonging the residence time of drugs in the absorbing tissues. For starch-based DDSs, high mucosal absorption of drugs could be achieved by coupling bioadhesive characteristics to carriers to prolong the residence time and/or widening the tight junction between cells to enhance the absorption at mucosal sites.

J. Chen et al., *Drug Delivery Applications of Starch Biopolymer Derivatives*
https://doi.org/10.1007/978-981-13-3657-7_5

5.1.1 Bioadhesion

When carriers contact with mucosa, adhesive bonds may be formed with the epithelial cell layer, the continuous mucus layer, or a combination of these two layers. The bioadhesion of starch MPs and NPs can be divided into two categories, i.e., (a) mucoadhesion between mucin and starch MPs and NPs; and (b) cell-specific receptor-mediated bioadhesion (Bie et al. 2016; Situ et al. 2015).

5.1.1.1 Mucoadhesion

The coupling of mucoadhesive characteristics to starch-based carriers has advantages such as efficient absorption and enhanced bioavailability of the drugs due to much more intimate contact with mucous layers. Mucoadhesive starch MPs and NPs can be tailored to adhere to any mucosa including those existing in the oral cavity, nasal cavity, GIT, vaginal, and eye. These DDSs can be used for the delivery of small drug molecules, peptides, and proteins, thus offering the possibilities of localized as well as systemic drug delivery (Table 5.1).

The mechanism of the adhesion of starch polymers to mucosa can be explained by the hydrogen bonding interactions between functional (such as hydroxyl and carboxyl) groups of starch and the glycoprotein network of the mucus layer or the glycoprotein chains attached to the epithelial cells (Illum et al. 1987; Lefnaoui and Moulai-Mostefa 2011, 2014; Mao et al. 2004; Surini et al. 2009). For achieving strong adhesive bonds, the establishment of intimate molecular contact between the starch polymers and glycoprotein chains is essential. Therefore, a basic requirement of the mucoadhesion of starch polymers is their ability to swell by absorbing water from the mucous layer to form a gel-like layer. The hydrophilic nature and swelling capacity of starch polymers play a vital role in the formation of mucoadhesion, and these attributes have to be retained and enhanced. For example, the mucoadhesion of phase-separated gelatin–starch blended hydrogels followed the order of hydrated starch > soluble starch > corn starch > gelatin, due to the increasing hydrophilicity of these three modified starches. The mucoadhesion of these hydrogels was shown to contribute to the extended drug release (Mallick et al. 2016). Moreover, in some cases, other mechanisms may be involved for the mucoadhesiveness of starch polymers, due to the particular feature of starch-based carriers. For example, owing to the negatively charged mucus layer, positively charged starch-based carriers favor for the formation of mucoadhesion (Marto et al. 2016; Saikia et al. 2016). Based on this knowledge, pregelatinized-modified starch nanocapsules with a ζ-potential of +33.6 mV have demonstrated to be potential for skin penetration for transdermal applications (Marto et al. 2016).

Increasing the content and the hydrophilicity of starch derivatives could help to improve the interactions between the matrices and the mucosal layer. Usually, higher swelling enhances mucoadhesion by interpenetration and adhesion of the polymers to the glycoprotein chains. However, when the excessive hydration of starch polymers

Table 5.1 Applications of the mucoadhesive property and the widening tight junction effect of starch-based carriers in transmucosal drug delivery system

Strategies	Therapeutics	Administration route	Modified starch used and dosage form	Comments	References
Mucoadhesion					
	5-Fluorouracil; Metronidazole	Oral	Assam Bora rice starch microspheres	Maximum mucoadhesive force was with large intestinal part Mean peak plasma concentration of drug loaded was lower than naked drugs	Ahmad et al. (2012a, b, 2013)
	Propranolol hydrochloride	Oral	Microspheres prepared from pregelatinized cassava starch succinate (PCSS) blending with Carbopol 974P or hydroxypropylmethyl cellulose (HPMC)	The microspheres of PCSS showed better adhesion at medium of pH 1.2 than that of pH 7.2 On gastric mucosa, addition of HPMC or Carbopol 974P reduced mucoadhesion; on intestine mucosa, they enhanced mucoadhesion Addition of HPMC or Carbopol 974P extended drug release, which was slower in pH 7.2 than pH 1.2	Surini et al. (2009)
	Metoprolol tartrate	Oral	Intragastric floating beads of tapioca starch blending with alginate beads	Addition of tapioca starch markedly improved sustained the release for 3–4 h (90% release within 45 min of alginate gel beads) Alginate–starch beads showed less than 50% mucoadhesion after 4 h	Biswas and Sahoo (2016)

(continued)

Table 5.1 (continued)

Strategies	Therapeutics	Administration route	Modified starch used and dosage form	Comments	References
	Ketoprofen	–	Hydrogels of gellan gum and retrograded starch blends	Dual crosslinking and starch retrograded by the cycled temperature (4 °C and 30 °C/16 days) favored the adhesiveness	Oliveira Cardoso et al. (2016)
	Miconazole nitrate	Oral cavity	Lozenges with bioadhesive layer that was composed of drum-dried waxy maize starch and Carbopol 980	Effective extended release of drugs was achieved after 15 min with a mean time of almost 3.5 h	Codd and Deasy (1998)
	Isosorbide dinitrates	Oral and buccal	Grafted starch microspheres using polyacrylic acid (PAA) and polymethyl methacrylic acid (PMMA)	Though the mucoadhesion in the case of PAA-grafted starch microspheres was higher than PMMA-grafted starch, the drug release exhibited relatively faster drug release	Vyas and Jain (1992)
	Curcumin	Injection	Aminated starch-coated magnetic nanoparticles (MNPs)	Mucoadhesion was higher at pH 5 than that at pH 7.4 At low pH, the release amount was more than that at high pH	Saikia et al. (2016)
	Isoniazid	Injection	Montmorillonite (MMT) containing MNPs coated with carboxymethyl starch–chitosan; Thiolated starch-coated MNPs/MMT nanocomposites	Addition of chitosan and MMT greatly enhanced mucoadhesive property; it was increased with the increased concentration of MMT The release of drug from MMT containing MNPs was found less than those of MMT free MNPs	Saikia et al. (2014, 2015)

(continued)

Table 5.1 (continued)

Strategies	Therapeutics	Administration route	Modified starch used and dosage form	Comments	References
Mucoadhesion + The widening tight junction effect					
	Insulin	Oral	Starch acetate (SA)-polyethylene glycol (PEG) micelles; molecular weight (*M*) of PEG: 500, 800, 1900	Moderate sustained release of insulin was observed at pH 6.8 and 7.4, with 60% and 80% of insulin released in 8 h Mucoadhesion of SA-PEG_{1900} was significantly higher than that of SA and SA-PEG_{800} and SA-PEG_{500}	Minimol et al. (2013)
	Insulin; sodium cromoglycate; rose bengal; melatonin; bovine serum albumin; domperidone CL	Nasal	Degradable starch microspheres (DSMs) or gel-crosslinked/spray-drying starch microspheres	The half-life of clearance was longer than the liquid and powder control formulations Partial separation of the tight junction of Caco-2 cells was observed, which coincided with the increased absorption rate of drugs Pulsatile or sustained release of drugs have been achieved and its bioavailability was high (e.g., absolute bioavailability of approx. 30% or 84.07%)	Björk and Edman (1988, 1990), Bjork et al. (1995), Edman et al. (1992), Illum et al. (1987), Mao et al. (2004), Soane et al. (1999), Witschi and Mrsny (1999), Yadav and Mote (2008)

(continued)

Table 5.1 (continued)

Strategies	Therapeutics	Administration route	Modified starch used and dosage form	Comments	References
	Influenza virus haemagglutinin; antigen and cytokines; insulin; desmopressin CL	Nasal; vaginal; sublingual	DSM and lysophosphati dylcholine (LPC)	Immune reaction (serum IgG and IgA antibody responses) and drug absorption (both in terms of peak plasma levels and bioavailability) were significantly enhanced by synergistically by DSM and LPC, e.g., a relative bioavailability of 3.6% of insulin; an absolute bioavailability of 13% of insulin	Critchley et al. (1994), Illum et al. (2001), Montgomery and Rafferty (1998), O'Hagan et al. (1993), Richardson et al. (1992)
	Insulin	Nasal	Starch nanoparticles prepared by different crosslinkers (epichlorohydrin, $POCl_3$) using emulsion or gel method + LPC	Particles prepared by emulsion method showed better and prolonged hypoglycemic action than gel method Mucoadhesion was shown to greatly enhance the absorption; drug release enhanced by using LPC	Jain et al. (2008)

occurred, a reduction in the strength of polymer–mucosa bounds was found due to the formation of slippery mucilage (Mortazavi and Smart 1993; Surini et al. 2009). Specifically, in the first step, even though the extensive swelling of starch-based carriers increases the contact surface between the polymer and the mucus, it also exposes repulsive forces facilitating the separation of the carriers from the mucosa. There were double effects of water excess in the mucoadhesiveness. First, the gel consistency could be reduced; second, water competed with the functional groups of oligosaccharides presented in surface mucosa for the formation of hydrogen bonds with polymers. Thus, a decrease in mucoadhesion might occur (Surini et al. 2009).

Decoration of starch with hydrophilic or polar functional groups by grafting or blend methods can help achieving a fast swelling and intimate contact between starch polymers and the mucous layer (Table 5.1). For instance, compared with PMMA, which has hydrophobic character, PAA-grafted starch MPs swelled considerably owing to the more hydrophilic nature of PAA, leading to apparent bioadhesion (Vyas and Jain 1992). Also, the addition of hydrophilic mucoadhesive polymers to starch or its derivatives is an easy way to enhance mucoadhesion of starch-based carriers by forming the secondary noncovalent bonds between the starch substrates and the mucin layer (Table 5.1). For instance, the mucoadhesive property of CMS was greatly enhanced when it combined with chitosan, due to the strong electrostatic interaction between the positively charged chitosan and the negatively charged glycocalyx (Rajan and Raj 2013). Thiolated starch-coated MNPs/montmorillonite (MMT) nanocomposites were found to display excellent mucoadhesiveness due to the increased interactions of disulfide bonds formed from thiol groups with mucus glycoproteins (Saikia et al. 2014). This improved mucoadhesion might also be ascribed to MMT (Saikia et al. 2014), the mucoadhesiveness of which was also verified in other studies (Güler et al. 2015; Saikia et al. 2015). MMT has a large surface area, negative charges, and hydrophilic properties, which can contribute to the development of London-van der Waals forces and hydrogen bonding between MMT and mucous cells. Thus, MMT exhibits excellent mucoadhesion (Campbell et al. 2008). Besides, when incorporating high amounts of MMT into starch-based carriers, another reason for enhanced mucoadhesion may be the reduced swelling capacity because of the co-crosslinking effect of MMT (Güler et al. 2015). The comparatively low swelling degree avoided the slippery mucilage of these carriers caused by the excessive hydration of starch polymers (Mortazavi and Smart 1993; Surini et al. 2009).

The mucoadhesiveness of starch polymers is a result of complex factors, and their mucoadhesive power was shown to be affected by the nature of starch polymers, the properties of carriers, and also by the nature of the surrounding medium (Table 5.2). Often, optimum preparation conditions for starch-based carriers are needed to achieve good mucoadhesion and extended drug release. Moreover, the factors that affect mucoadhesion should be all considered to avoid the failure of mucoadhesive starch-based DDSs. For example, applying the same mucoadhesive starch polymers to different DDSs might lead to different performances. Crosslinked CMS MPs have been successfully used for nasal or pulmonary drug delivery, but their high hydration and swelling capacity in the GIT fluids with the resulted poor mucoadhesion, limit their application to oral drug delivery (Lemieux et al. 2015).

Table 5.2 Factors affecting the performance of starch-based bioadhesive drug delivery system

Factors	Comments [Refs.]
Hydrophilicity/hydrophobicity	The hydrophilic nature of modified starch was necessary (Mallick et al. 2016; Minimol et al. 2013), higher hydrophilic of modified starch, higher mucoadhesion (Mallick et al. 2016)
Molecular weight (*M*)	High *M* polymers of starch-based matrix increased the polymer/mucus interpenetration (Björk and Edman 1990; Minimol et al. 2013) No optimum *M* of starch is found
Concentration	Mucoadhesion was increased with the increased concentration of starch (Ahmad et al. 2012a, b, 2013; Lefnaoui and Moulai-Mostefa 2011, 2014; Yadav and Mote 2008) When blending with other polymers, optimum concentration is required for best bioadhesion; high concentration may result in coiling of polymer molecules and hence reduced the flexibility of the polymeric chains E.g., maximum capacity of adhesion with a concentration of 0.5% of kappa carrageenan to pregelatinized starch gel was found, beyond this concentration, the adhesion decreased (Lefnaoui and Moulai-Mostefa 2011, 2014)
Flexibility of polymer chains	Required for interpenetration and entanglement Medium crosslinked degree showed a better and prolonged drug effect compared to the low and high degree of crosslinked nanoparticles (Güler et al. 2015; Jain et al. 2008) High crosslinked polymers: mobility of individual polymer chains and the effective length of the chain which can penetrate into the mucus layer decreased, which led to the decreased bioadhesion strength (Oliveira Cardoso et al. 2016)
Properties of dosage forms	
Particle size	Low particle size and high surface to volume ratio contributed to strong adhesion to the mucosal surface (Saikia et al. 2014)
Swelling	The swelling capacity depends on time, polymer concentration, crosslinking degree and degree of substitution (DS), pH and chemical structure of modified starch, the amount of water at mucosa environment A low swelling degree was required to prolong mucoadhesion and the greater rate of swelling, then shorter the duration of adhesion (Güler et al. 2015; Surini et al. 2009) Swelling was increased with an increase in time (Saikia et al. 2016) Increasing crosslinking degree reduced the flexibility and polymer relaxation of starch thus reducing swelling capacity (Björk and Edman 1988, 1990; Edman et al. 1992; Saikia et al. 2014, 2016) Increased DS of hydrophilic functional groups to starch led to higher water holding content due to the enhanced hydrophilicity of starch (Lemieux et al. 2015) Ionized and protonated of functional groups of starch in different pH results in different interactions, thus polymer relaxation: for carboxyl and hydroxyl modified starch, swelling was more at high pH compared to low pH (Güler et al. 2015; Lemieux et al. 2015; Saikia et al. 2014, 2015); for cationic starch, swelling percentage was more at low pH than high pH (Saikia et al. 2016) The amount of water at mucosa environment should be not too much or little for carriers to swell (Lemieux et al. 2015), take nasal administration as an example, much microparticles in nasal exceeded the moisture available for swelling, then a proportion remained dry and the drug could not be released (Jain et al. 2008)

(continued)

Table 5.2 (continued)

Factors	Comments [Refs.]
Environmentally related factors	
pH	The change of surface charge on polymer and degree of hydration are related to the change of pH Protonated carboxyl or hydroxyl groups form more hydrogen bonds with mucin strains than the ionized groups (Ahmad et al. 2012a, b, 2013; Biswas and Sahoo 2016; Lemieux et al. 2015; Surini et al. 2009) The interaction between charged modified starches and negatively charged mucus layer: positive charge starch-based vectors favor for mucoadhesion (Marto et al. 2016; Saikia et al. 2016)
Initial contact sites	The drug release behavior and clearance of starch-based carriers were affected (Codd and Deasy 1998), e.g., compared to the ciliated region of the nose which clearance was rapid, particles deposited within the anterior region of the nasal cavity can remain static for up to 2 h (Newman 1987)

Among all the environmental factors affecting starch-based mucoadhesive DDSs, the most important one is pH. For oral site-specific and tumor-targeting drug delivery, starch-based mucoadhesive DDSs can be particularly designed to be responsive to pH. The degree of hydration of polymers can be dependent on its chemical structure and the environmental pH (Table 5.2). For example, ionic starch polymers have been applied to achieve pH-sensitive mucoadhesion. Cationic groups such as amine groups had to be incorporated to starch derivatives to allow enhanced mucoadhesion at acidic tumor microenvironments. Regarding this, these groups could be protonated to acquire polymers with a high positive charge density, and thus to facilitate the interactions of the polymer with the negative cell mucosa in tumor acidic microenvironments (Saikia et al. 2016). Besides, anionic groups can be incorporated into starch to achieve pH-responsive mucoadhesion of starch-based oral site-specific DDSs. Taking CMS MPs as an example (Lemieux et al. 2015), under acidic conditions, the lower water uptake of the CMS MPs and a higher amount of hydrogen bonding between carboxylic groups and mucus glycoprotein represented a favorable condition for mucoadhesion. Nonetheless, under high pH conditions, the lack of intestinal mucoadhesion of the CMS MPs may be due to the repellency of the carboxylate groups on the starch by the negatively charged mucin chains, and repulsive forces that facilitated the separation of the MPs from the intestinal mucosa caused by the excessive hydration. Thus, the mucoadhesion was much higher in the gastric mucosa than intestinal mucosa (Lemieux et al. 2015). Besides ionic starch polymers, the hydroxyl groups and ester bonds of starch polymers can also be protonated to small extents along the GIT, making these starch polymers more mucoadhesive to the gastric mucosa (Surini et al. 2009) or large intestinal regions (Ahmad et al. 2012b).

5.1.1.2 Cell-Specific Receptor-Mediated Bioadhesion

The phenomenon of mucoadhesion, however, is unpredictable due to the varying turnover time and the diverse composition of mucus and disease conditions. The lack of specificity in adhesion to mucous tissue limits drug delivery through this technique. To address this issue, specific mucosal surfaces have been investigated to be targeted by site-specific chemical agents that are anchored onto starch polymeric DDSs to enhance the therapeutic benefits of drugs. Up to date, there has been limited research on cell-specific receptor-mediated bioadhesion of starch-based DDSs. The specific targeting of drugs to absorption sites can be achieved by anchoring plant lectins, bacterial adhesins, and antibodies on the surface of starch-based carriers (Vasir et al. 2003). A glycoprotein, concanavalin A (Con A), which is one of the most abundant lectins that could bind to the mannose of glycoproteins on the surface of cells, has been successfully conjugated to starch polymers, leading to the formation of starch polymers–glycoprotein complex (Bie et al. 2016; Situ et al. 2015). The conjugation of Con A to the surface of SA films could increase the residence time of film-coated MPs in the colon and enhance the absorption efficiency of bioactive molecules (insulin) through bioadhesion to the surface of colonic mucosa cells (Bie et al. 2016; Situ et al. 2015).

5.1.2 The Widening Tight Junction Effect

When it comes to the paracellular route of drug permeation across the epithelial membranes, the bioavailability of starch polymeric DDSs can also be enhanced by the widening tight junction effect, combined with prolonging the contact time with mucosa by improving its bioadhesive properties (Table 5.1). The mechanism of the action of starch particulate carriers on tight junctions can be associated with the dehydration of epithelial mucosa due to the swelling capacity of starch particulate carriers. Then, a reversible "shrinkage" of the cells occurs, giving a physical separation of the intercellular junctions and facilitating passage of drugs through the paracellular route (Boddupalli et al. 2010). Theoretically, any bioadhesive starch-based DDSs that have hydrophilicity and can swell when contacting with mucosa within a short time has the ability to widen tight junction between mucosa cells.

One of the most typical starch-based carriers for mucoadhesive DDSs are starch MPs, especially DSMs for nasal drug delivery (Table 5.1). These DSMs have been prepared by emulsion polymerization technique where starch was crosslinked with epichlorohydrin (Björk and Edman 1988, 1990; Critchley et al. 1994; Edman et al. 1992; Illum et al. 1987). Other bioadhesive starch MPs using different crosslinkers or different preparation methods have also been reported such as inverse emulsion method using glutaraldehyde (Momeni and Mohammadi 2009), and spray-drying (Witschi and Mrsny 1999). When these starch MPs were used for nasal administration, the mucoadhesion could allow these microspheres to be cleared more slowly from the nasal cavity, thereby increasing the contact time of the carriers with the nasal mucosa (Mao et al. 2004; Soane et al. 1999). For instance, the half-lifetime of the clearance of DSMs from nasal cavity was found to be in the order of 240 min as compared to 15 min for the liquid and powder control formulations (Illum et al. 1987). Also, the increased absorption rate of drugs was reported to be achieved by widening the tight junctions between cells (Björk and Edman 1988, 1990; Bjork et al. 1995; Edman et al. 1992). When using DSMs to delivery peptide drugs (insulin), the insulin transport increased and a partial separation of the tight junction of monolayers of human intestinal epithelial (Caco-2) cells was observed. After 15 min, insulin transport was further enhanced and the tight junctions remained separated (Bjork et al. 1995). The DSMs induced a pulsed delivery of insulin across the epithelium that lasted for 1–2 h, which was correlated with a reversible appearance of focal dilatations in the tight junctions between the epithelial cells (Bjork et al. 1995). Thus, it can be implied that, under the combined effects of bioadhesion and the widening tight junction effect, starch MPs can significantly improve the absorption of drugs, both regarding peak plasma levels and bioavailability, e.g., a rapid absorption rate ($T_{max} = 7.8$ min) and a high absolute bioavailability (84.07%) of drugs (melatonin) by intranasal administration using starch MPs (Mao et al. 2004).

Despite the good performance of DSMs for enhancing absorption, the combination DSMs with absorption enhancers is still needed to improve the bioavailability of drugs in the mucosal application. For example, the combination of starch MPs and lysophosphatidylcholine (LPC), which is the most common absorption enhancer, showed a significant improvement in bioavailability in comparison with the only microsphere delivery and the LPC solution formulation (Critchley et al. 1994; Richardson et al. 1992). The enhancement of absorption by DSMs did not exceed the ability of absorption enhancers, which will limit its practical application. More efforts are need regarding this concern. In recent years, results indicated that the effect of starch particulate carriers on tight junction can be improved through modulating biomolecules and physicochemical factors that regulate tight junction structure and permeability, such as tight junction proteins (Reinhold and Rittner 2017) and intracellular calcium concentrations (Madhav et al. 2009). For example, it was reported that a higher DS of the CMS MPs led to not only dehydration of epithelial mucosa but also the increased widening tight junction effect. The explanation was that the increased calcium-binding capacity of CMS with a high DS might deplete extracellular Ca^{2+}, which was necessary to maintain the tight junction (Lemieux et al. 2015). Moreover, a temporary widening of the tight junctions in a monolayer of human epithelial cells was seen in the presence of PEGylated SA NPs, and the staining intensity of the tight junction proteins ZO-1 was weaker at cell–cell contact sites because of the tight junction disassembly and paracellular permeability changes induced by PEGylated SA NPs (Minimol et al. 2013).

The intimate contact of the carriers with mucosa may hypothetically provoke perturbations to the epithelial barrier including the structural separation of tight junction, which will have an extremely adverse effect and activate the death pathways of the cells. Thus, the understanding of the interactions between starch-based carriers with the mucosa epithelial barrier could be highly significant, although very few studies concerning this issue have been reported so far. Moreover, there is no trend of the effect of starch-based carriers on the mucosal cells. Despite less effective enhancement of absorption, DSMs have a huge advantage over many absorption-enhancing agents, i.e., they do not compromise the vital barrier function of the epithelium (Anderberg et al. 1992). Nevertheless, when evaluating the influence of

CMS MPs on the tight junction dynamic (barrier) function of gastric and intestinal cell monolayers, there was a big difference between the effect on the gastric and intestinal cell monolayers (Lemieux et al. 2015). The tight junctions of gastric cell monolayers were not influenced in the presence of CMS MPs, whereas that of intestinal cell monolayers decreased with a higher DS but recovered to its initial value at about 15 h after the treatment. This difference was due to the presence of the mucus cluster at the surface of the gastric cell monolayers, which could protect the tight junctions of the monolayer from the action of CMS MPs to maintain their initial barrier function (Lemieux et al. 2011). Thus, combining with its higher mucoadhesion at gastric mucosa, this feature makes these CMS MPs particularly suitable for gastric retentive drug formulations.

5.2 Starch-Based Target-Specific DDSs

Improvements in target-to-nontarget concentration ratios, increased drug residence at the targeted site, and improved cellular uptake and intracellular stability are some of the major reasons for the great emphasis on the use of target-specific DDSs. Yet, in the application of target-specific drug delivery, given the potential toxicity of drug-loaded carriers, tissue selectivity is a major issue, especially for the solid tumor treatment, because of the small therapeutic window of chemotherapeutic agents and the steep dose-response curve (Frei and Canellos 1980). Phagocytosis of particles can be initiated by opsonization (e.g., the interactions with opsonin proteins) through their blood circulation (Aderem and Underhill 1999), leading to their rapid clearance or unwanted immune response by the RES. Thus, there could be a high risk of off-target delivery for the macrophage-evading target-specific systems. In this section, to achieve the biocompatibility of starch-based carriers and the improvement in the targeting efficiency of starch-based nanocarriers, the plasma hematocompatibility, the unwanted recognition by the RES, and the targeting strategies (passive, active and physical targeting) for macrophage-evading, target-specific, starch-based DDSs will be discussed. Nevertheless, the targeting delivery of immunoreactive substances to the immune system in the human body (RES cells and mucosal M cells) can be an effective way to enhance the immune response for the treatment of intracellular infections. Thus, the RES cells and M cells targeting of starch-based carriers will be also focused in this section.

5.2.1 Plasma Hematocompatibility

Understanding NP hematocompatibility is an important step in the initial characterization of nanomaterials. On the behalf of good hematocompatibility, an essential requirement for the intravenous use of NPs is their long circulation time in the blood in a highly stable manner to avoid hemolysis, thrombogenicity, or the activation of phagocytosis by immunes (Dobrovolskaia et al. 2008). Such NPs must possess "stealth" properties to maintain their stability in body fluids and minimize the protein absorption in the blood. Starch has been investigated to modify the surface of its NPs to achieve "stealth" properties. The water barrier created by strong hydrogen bonding of hydrophilic starch groups with surrounding water molecules can resist proteins from approaching the surface of these NPs. One classical example is starch-coated MNPs. The surface coating of MNPs by starch has been used to avoid "agglomeration", which is the biggest problem in the synthesis of iron oxide NPs (Kim et al. 2001).

To improve plasma hematocompatibility, the hydrophilicity of starch-based materials is necessary. It was found that increasing the crosslinker concentration tended to enhance the hydrophobicity of the NPs, which, in turn, resulted in an increased thrombogenicity as evidenced by the observed higher values of hemolysis of crosslinked starch-impregnated MNPs (Jain and Bajpai 2013; Likhitkar and Bajpai 2012). Besides, the size and surface charge of starch NPs are important factors. Usually, the blood half-life of nanocarriers was shorter for the increasing size of the particles (Roohi et al. 2012). After injected into the blood serum, except few carriers with positive charges, e.g., cationic starch polymers for DNA delivery, most starch-based nanocarriers are negatively charged at physiological pH (pH 7.2), which favors for good biocompatibility and nontoxicity of starch nanocarriers (Lin et al. 2010). Yet, this surface feature of starch NPs may lead to the electrostatic interaction between proteins and starch nanocarriers, because most of the proteins are above their isoelectric point (Höök et al. 1998). Thus, a certain range of surface charge that the electrostatic repulsion between nanoparticles can be high enough to avoid agglomeration while at the same time not high enough to combine with proteins. A nearly neutral charge will favor the non-protein absorption and the RES recognition (Ye et al. 2016). However, it was reported that even though the surface charge of NPs is close to neutral [the ζ-potential of starch-coated MNPs was -4.15 mV (Ruge et al. 2012)], those NPs can still absorb proteins and increase their chance to be cleared by RESs. These studies prove the multifactorial and complicated interactions between nanocarriers and proteins/RES cells. Moreover, the uptake by the RES is unfortunately unavoidable in most cases.

Despite the good "stealth" properties of starch, in some cases, grafting PEG onto starch is still needed to limit the protein adsorption to a certain degree (Zhang et al. 2013a). However, PEG has been reported to have several fatal drawbacks in clinical applications, such as the accelerated blood clearance of NPs after a second dose, resulting from the generation of anti-PEG antibodies in the immune system (Armstrong et al. 2007; Zhang et al. 2013c). There is a desperate need for a substituted polymer of PEG. HES has the potential as a natural, nontoxic alternative for PEG because of similar protein repellent characteristics (Freichels et al. 2013; Narayanan et al. 2015; Noga et al. 2013, 2014; Winzen et al. 2015). No unfavorable accumulation of HES has been observed in the liver or spleen (Hoffmann et al. 2013). Also, a higher dosage (20–36 ml/kg) of HES than recommended (maximum of 20 ml/kg) showed no significant increase in bleeding or organ toxicity (Narayanan et al. 2015). All this makes HES a good candidate to replace PEG. When using HES NPs as a delivery carrier, higher stability of NPs in the blood serum and no hemolysis of red blood cells were found (Narayanan et al. 2015). The HES-decoration of PEI reduced the cytotoxicity of PEI by the effect of steric hindrance against protein-induced aggregation (Noga et al. 2013, 2014).

Decoration starch-based nanocarriers with functional groups could be a good way to enhance "stealth" properties of starch further. Recently, a starch derivative has been prepared with the incorporation of a superhydrophilic moiety of zwitterionic sulfobetaine and a hydrophobic segment of octane (Ye et al. 2016). This modified starch could self-assemble into "stealth" micelles. It was found that the micelles modified with only octane showed significant protein adsorption leading to the aggregation of the micelles. However, the micelles modified with both sulfobetaine and octane groups were stable in 10% BSA and 10% human serum solutions, without obvious size changes at the end of 11-day incubation. These results suggested lower macrophage activation of the micelles due to the strong hydration layer via electrostatic interactions, eventually extending their circulation time in the blood (Ye et al. 2016). Moreover, compared with the micelles only modified with octane (ζ-potential: −24.9 mV), the nearly neutral charge of the micelles with octane and zwitterionic modifications (ζ-potential: −2.4 mV) could attribute the longer blood stability time and lower macrophage activation (Ye et al. 2016). The enhanced "stealth" effect by the functionalization of starch relies on the modulation of physicochemical properties (e.g., composition, surface charge, hydrophobicity) of carriers. The greater the improvement in the physicochemical properties of starch-based carriers, the more enhanced "stealth" effect has been observed. For instance, due to the similar parameters of functionalized HES NPs to HES NPs, HES nanocapsules (size: 275 nm; ζ-potential: −34 mV) functionalized with carboxyl groups (size: 200 nm; ζ-potential: −34 mV) and amino groups (size: 256 nm; ζ-potential: −27 mV) had a minor effect on the adsorption and interaction with human serum albumin (HSA) (soft protein corona) and apolipoprotein (Apo)A-I (hard protein corona) (Winzen et al. 2015).

There is still some debate on the benefits or disadvantages of protein binding. On one hand, these protein-marked NPs can increase the chance of clearance by the RES, which will not only decrease the efficiency of targeting, especially in the case of passive targeting, but also may induce the immune response of the body. On the other hand, specifically bound proteins can also help target or direct the nanoparticle toward a particular pathway or area of the body. For example, it was observed that lung surfactant-associated proteins A and D that have immunostimulatory activity, could enhance the alveolar macrophage uptake of the hydrophilic starch MNPs (Ruge et al. 2012). Furthermore, the absorption of proteins with NPs could also contribute to higher cellular uptake. When pteroic acid conjugated polyvinyl alcohol was grafted onto propyl starch NPs, higher adsorption of the model protein (80% BSA) on the formulated NPs contributed to the enhanced cellular uptake of NPs as compared to polyvinyl alcohol (Jain et al. 2011). Therefore, engineering a particle that can specifically bind certain proteins of interest for targeting purposes, or engineering particles with these proteins already bound to the particle, can greatly enhance the ability to develop directed NPs as drug candidates.

5.2.2 *Macrophage-Evading Target-Specific System*

Targeting delivery of bioactive compounds at living organisms with minimal toxicity has attracted growing interests in the development of starch-based colloidal carriers. To improve the targeting efficiency of starch-based DDSs, starch-based particles can be modified for targeting at tumor cells, by different approaches including (a) passive targeting, (b) active targeting, and (c) physical targeting by EMF. These different approaches are discussed below in detail.

5.2.2.1 Passive Targeting

Passive targeting relies on the properties of DDSs and the disease pathology to preferentially accumulate drugs at the site of interest and avoid nonspecific distribution. The development of long-circulating NPs allows many starch-based platforms to exploit structural abnormalities in the vasculature of particular pathologies (Hallenbeck et al. 1986). In this regard, starch NPs, e.g., propyl starch NPs (Dandekar et al. 2012), by virtue of the EPR effect (Chap. 4, Fig. 4.2), have been found to have an antitumor effect in the tumor mass by passive targeting of the incorporated drugs toward the tumor cells. The nonspecific accumulation, or passive targeting, has been demonstrated with NPs ranging from 10 to 500 nm in diameter (Table 5.3), which allows these NPs preferentially being accumulated near the tumor mass via the EPR effect and be uptaken by cells via endocytosis. Nevertheless, large NPs are more likely to be cleared by the RES cells and have short blood half-life (Roohi et al. 2012). Thus, based on the preferred size range (50–150 nm) of cancer cell uptake (Satchi-Fainaro et al. 2006) and the size (approximately 200 nm in width) of the inter-endothelial cell slits of the spleen (Moghimi et al. 2001), starch NPs of comparatively small sizes (<200 nm) is highly appreciated. For example, HES–PEI copolymers of 70–80 nm in size markedly reduced the nonspecific lung transcription of the complexes to maintain gene expression in the tumor (Noga et al. 2013, 2014).

Among all methods and strategies for the construction of starch-based nanocarriers, drug–HES conjugates represent an interesting example to be addressed. Drug–HES conjugates could significantly improve the drug stability as the drugs were shielded by HES. When drug–HES conjugates were administered, a high antitumor efficacy was found due to the EPR effect (Kumar et al. 2014; Li et al. 2014a, 2016), and their half-lifetime in plasma was 30-fold and 40-fold higher in comparison to that of unconjugated drugs (Li et al. 2014a). However, the reasons for the higher antitumor efficacy of drug–HES conjugates are the specificity to the drug type. For example, the superior effect of phloroglucinol-HES could be due to the hydrophobicity of the grafted biopolymer, which may enhance the adhesion and adsorption onto the surfaces of cancer cells (Kumari et al. 2010). In contrast, the conjugation of HES to 10-hydroxy camptothecin (10-HCPT) assisted 10-HCPT to escape the fate of hydrolysis to its carboxylate form, which was hard to enter the cell because of the higher polarity. Thus, this 10-HCPT–HES conjugates had an excellent cell pen-

Table 5.3 Strategies developed to improve the performance of starch-based targeting delivery

Target	Strategies to achieve targeting	Parameter	Strategies to enhance transport [Refs.]	Examples [Refs.]
Macrophage-evading target-specific drug delivery systems (DDSs)				
Passive targeting				
Tumor cells	Enhanced permeability and retention (EPR)	Size	Uniform size distributions and range of mean sizes reported from 10 to 500 nm	E.g., hydroxyethyl starch (HES)-methotrexate conjugates: 15.2 nm (Goszczynski et al. 2014); starch-*g*-polyethyleneimine (PEI)/DNA complexes: 70–100 nm (Yamada et al. 2014); phloroglucinol-starch conjugate nanocomposites: 493.3 nm (Kumar et al. 2014)
			Smaller size favors for cellular uptake and targeting efficiency	Folic acid (FA)-conjugated HES nanocapsules with smaller size (174 nm) showed a quantitatively better uptake than those with an average size of 307 nm (Baier et al. 2012)
		Hydrophobicity/hydrophilicity	Mostly hydrophilic, except in rare cases, hydrophobic modifications used	Hydrophilic modifications, e.g., polymethacrylic acid-polysorbate 80–starch (Shalviri et al. 2012), HES-drug complex (Li et al. 2014a; Noga et al. 2013, 2014), HES-g-polylactide acid copolymers (Liu et al. 2013) Hydrophobic modifications, e.g., propyl starch (Dandekar et al. 2012), acetylated starch (Xiao et al. 2016)
			A better balance between hydrophobicity and hydrophilicity favors for cell uptake	Acetylated starch nanocrystals exhibited slower drug-release rates and higher antitumor activity than raw starch nanocrystals. However, with the increased degree of substitution (DS) from 0.04 to 0.14, the cell cytotoxicity increased (Xiao et al. 2016)

(continued)

Table 5.3 (continued)

Target	Strategies to achieve targeting	Parameter	Strategies to enhance transport [Refs.]	Examples [Refs.]
		Surface charge	Mostly negatively charge, only a few of them with a positive charge	Zeta potential, e.g., HES-methotrexate conjugates: −27.7 mV (Goszczynski et al. 2014); quaternized starch/small interfering RNA: +27.7 mV The zeta potential values for the FA tagged aminated starch/ZnO coated iron oxide nanoparticles changed from 38 to 43 mV (Saikia et al. 2017)
			Cationic starch polymers favor the cellular uptake, yet high charges induce high cytotoxicity to cells (Constantin et al. 2003; Engelberth et al. 2015; Sieradzki et al. 2014)	The positive charged ZnO/aminated starch MNPs might be significantly uptaken by tumor cells since the cell membrane contains large negative charges (Saikia et al. 2017)
			A nearly neutral charge is recommended	Longer blood stability time and lower macrophage activation of micelle with octane and zwitterionic modifications (ζ-potential: −2.4 mV) compared with micelle only modified with octane (ζ-potential: −24.9 mV) (Ye et al. 2016)
Active targeting				
Tumor cells	Cell-specific ligand	–	FA (Baier et al. 2012; Saikia et al. 2017; Wang et al. 2014; Xiao et al. 2006) targeting to FRα-mediated cells, mostly used targeting ligands	Cellular uptake of the FA conjugated chitosan/dialdehyde starch/mesoporous silica nanoparticles in folate receptor-overexpressing HeLa cells is much higher than that of non-folate acid receptor-overexpressing A549 cells (Wang et al. 2014)

(continued)

Table 5.3 (continued)

Target	Strategies to achieve targeting	Parameter	Strategies to enhance transport [Refs.]	Examples [Refs.]
			Pteroic acid (Thiele et al. 2011), hyaluronic acid (HA) (Kwag et al. 2014; Li et al. 2015) for targeting to cancer cells Polypeptide (AGKGTPSLETTP peptide) targeting to human hepatocellular carcinoma cell line BEL-7402 (Yang et al. 2009) Polysorbate 80 targeting brain metastases of breast cancer (Li et al. 2014b)	Antitumor effect of nanocarriers assembled with amphiphilic starch micelles and HA nanoparticles in tumor cell lines PC-3, A549, and HeLa cells: significantly higher antitumor efficiency in reducing tumor size compared with free drugs; A549 cells were not very sensitive to drug; while PC-3 cells suffered more toxicity of drugs (Li et al. 2015)
Magnetic targeting				
Tumor cells	External magnetic force (EMF)	Strong magnetic force response	Improve the targeting efficiency	In the presence of EMF, ortho-carborane encapsulated starch-coated magnetic nanocomposites showed high targeted delivery and the boron concentrations in tumor reached a high value of 51.4 μg/g tumor with tumor/normal tissue of around 10:1 (American Psychiatric 1988)
		Starch coating	Ensures homogenous distribution and reduced agglomeration, increased biocompatibility (Bergemann et al. 1999; Darroudi et al. 2014; Kim et al. 2003; Lu et al. 2013) Hydrophilicity of starch is necessary Need to be further modified starch with other hydrophilic polymers	Grafting starch-coated MNPs with PEG (Cole et al. 2011a, b), PEG and heparin (Zhang et al. 2013b, b, c; Zhang et al. 2014), and β-glucosidase (Jie et al. 2013) to further inhibit reticuloendothelial system (RES) E.g., reductions in liver (~12-fold) and spleen (~2.5-fold) PEG-starch-coated magnetic nanoparticles (MNPs) concentrations at 1 h compared to only starch-coated MNPs; and 15-fold improvement in targeting efficiency over previously targeted only starch-coated MNPs (Cole et al. 2011a)

(continued)

Table 5.3 (continued)

Target	Strategies to achieve targeting	Parameter	Strategies to enhance transport [Refs.]	Examples [Refs.]
DDSs that targeting the immune system				
Immune cells	Reticuloendothelial system (RES)	Size	Contrast to EPR, larger size, more chance to be recognized by RES	The blood half-life was shorter for larger particles of starch-coated magnetic nanoparticles (Roohi et al. 2012)
		Hydrophobicity/hydrophilicity	Mostly hydrophobic modification, but also hydrophilic modification existed	Hydrophobicity modification, e.g., acetylated starch microparticles (DS $\geq$ 2) with a typical RES distribution (Baillie et al. 1987; Laakso et al. 1986; Stjarnkvist et al. 1989, 1991; Thakore et al. 2013); Hydrophilic modification, e.g., HES nanocapsules (Fichter et al. 2013)
			More transportation of hydrophobic nanoparticles than hydrophilic nanoparticles	Polypeptide (poly-1-tyrosine) was used to enhance acryloylated HES microspheres hydrophobicity to improving uptake of M cells (Chadha and DeLuca 1998)
		Cell-specific ligands	Lectin coupled (oligo) mannose ligands for targeting of macrophage and dendritic cells	Uptake of nanocapsules was highly enhanced by the mannose on the surface of nanocapsules; and compared with a-D-mannopyranosylphenyl isothiocyanate, 3-*O*-(a-D-mannopyranosyl)-D-mannopyranose interacted less with human serum protein; but both better than camannose, thus had a better binding with lectin (Freichels et al. 2013)

(continued)

Table 5.3 (continued)

Target	Strategies to achieve targeting	Parameter	Strategies to enhance transport [Refs.]	Examples [Refs.]
		Combination with adjuvants	Monophosphoryl lipid A (MPLA) to enhance uptake and maturation of DCs Anti-CD40, anti-DEC205 targeting DCs through CD205 and CD40 receptors	HES microcapsules equipped with MPLA, anti-CD40, and anti-DEC205 induced the secretion of opsonins (Fichter et al. 2015; Pietrzak-Nguyen et al. 2014)
	Membranous (M) cells	Size	The optimal size range for effective internalization by M cells: 1–10 μm	E.g., acetylated HES microspheres: 1.7–10.2 μm (Chadha and DeLuca 1998); poly(acrylic starch) microspheres: 3.2 ± 0.4 μm (Sturesson and Degling Wikingsson 2000)
		Hydrophobicity/hydrophilicity	More transportation of hydrophobic nanoparticles than hydrophilic nanoparticles	Polyacrylation of starch (Larhed et al. 2004; Strindelius et al. 2004; Sturesson and Degling Wikingsson 2000) has been used to increase the hydrophobicity of starch-based carriers

etrating property and lactone stability (Li et al. 2014a, 2016). This suggests the drug properties have a huge impact on the tumor inhabitation and should be considered for the preparation of HES-drug conjugates.

As discussed in Sects. 2.2 and 5.2.1, it is evident that physicochemical properties of starch NPs (e.g., hydrophobicity, size, and surface characteristics) play important roles in determining its interactions with cells and biomacromolecules. General trends have emerged to guide the design of starch NPs to avoid recognition by the RES and to improve targeting efficiency of starch-based DDSs (Table 5.3). Taking size as an example, there is the trend that comparatively small size of particles favors the high stability of blood circulation and cellular uptake, thus high targeting efficiency. For example, different sizes resulted from two different preparation methods of PMAA-Polysorbate 80 (PS80)-*g*-starch NPs, namely preformed NPs (crosslinked PMAA-PS80-*g*-Starch) and self-assembled NPs (noncrosslinked PMAA-PS80-*g*-Starch) (Shalviri et al. 2012). The mean sizes of the preformed and self-assembled NPs were 137 and 62 nm, respectively, while the ζ-potentials were similar (-35 and -38 mV, respectively). Due to the size difference, the amount of the preformed NPs accumulated in the tumor appeared to be substantially lower than that of self-assembled NPs (Shalviri et al. 2012). Furthermore, a big difference in the distribution of PMAA-PS80-*g*-starch NPs in mice was observed. The preformed NPs exhibited substantially higher levels of accumulation in liver, spleen, lungs, heart, and intestine, being largely excreted via the hepatobiliary route. However, significantly higher levels of self-assembled NPs could be detected in kidney and tumor, being excreted via the renal route (Shalviri et al. 2012). These results verify that a large particle size of NPs increases their risk of being cleared by RESs, thus a lower targeting efficiency and cellular uptake were expected.

Yet, due to the complexity of the biological interaction, the effect of physicochemical parameters on the fate of NPs often vary with different polymeric particles and diffident interactions. The hydrophilic modification of starch is needed to extend the blood circulation time and improve the stability of its carriers (Jain and Bajpai 2013; Likhitkar and Bajpai 2012). Therefore, starch-based carriers with such features can attribute to the high targeting efficiency of passive targeting by the inhabitation of the RES recognition. Nonetheless, HES nanocapsules have been found to be predominantly incorporated by KCs, and drug-loaded HES nanocapsules reduced the release of inflammatory cytokines significantly in response to the bacterial endotoxin stimulation (Fichter et al. 2013). Moreover, it was noticed that the introduction of hydrophobic groups to starch could also allow modified starch-based DDSs to be passively targeted to tumor cells via EPR (Dandekar et al. 2012). For example, hydrophobic acetylated starch nanocrystals exhibited higher antitumor activity than raw starch nanocrystals. However, with the increased DS from 0.04 to 0.14, the cell cytotoxicity increased (Xiao et al. 2016). Thus, a greater balance of surface hydrophilicity/hydrophobicity is required to achieve a high cellular uptake and low cytotoxicity. Therefore, when to design a specific starch-based DDS, more work is needed to study its own particular features, despite the general trends of target-specific DDS.

5.2.2.2 Active Targeting

Active targeting is important to minimize nonspecific interactions and the uptake of NPs by nontargeted cells. Against solid tumors as an example, several strategies can be adopted for the surface modification of starch-based nanocarriers for effective targeted delivery to tumor cells or the endothelial cells of the tumor blood vessels (Table 5.3). Since tumor cells are rapidly proliferating, they overexpress certain receptors for the enhanced uptake of nutrients, such as FA and sugars. Active targeting to the disease site relies on the coupling of a specific ligand on the carrier surface that will be recognized by the cells present at the disease site. Up to now, the research on the active targeting of starch-based DDSs is rare.

FA is the most common ligand to achieve active targeting. Modification the surface of starch-based nanocarriers with FA may have advantages for the intracellular delivery of sensitive bioactives, such as protein and DNA since particles internalized by folate receptor-mediated endocytosis might escape trafficking into lysosomes and often are retained in endocytic compartments or released directly into the cytoplasm (Kelemen 2006). The cellular uptake of the FA-conjugated starch NPs in FA receptor-overexpressing cells was much higher than in normal cells (Baier et al. 2012; Saikia et al. 2017; Wang et al. 2014; Xiao et al. 2006). This was exampled by the observation that a very low cellular uptake was observed with nonfunctionalized HES nanocapsules. Nevertheless, the conjugation of HES with FA enabled a specific cellular uptake into HeLa cells, suggesting a folate receptor α-mediated uptake in HeLa cells (Baier et al. 2012). The other components in FA-functionalized nanocarriers may also affect the cellular uptake efficiency. For example, the incorporation of ZnO into the FA-tagged aminated starch MNPs increased the tumor cellular absorption, and this increase became more significantly with a higher content of ZnO (Saikia et al. 2017). This can be explained by the increased positive surface charge of these NPs with the increased ZnO content. At the cancer cell pH, the presence of cationic $ZnOH^+$ further furnished more positive surface charge to the drug-loaded NPs (Saikia et al. 2017).

Other compounds such as pteroic acid (Thiele et al. 2011), polypeptide (Yang et al. 2009), HA (Kwag et al. 2014; Li et al. 2015) and PS80 (Li et al. 2014b) have also been applied to starch-based nanocarriers to improve the targeting delivery of drugs (Table 5.3). For example, PMAA-PS80-*g*-starch NPs can specifically bind ApoE, and selectively target to the brain across the blood–brain barrier (Shalviri et al. 2012). PS80-coated NPs recruited ApoE in the plasma, mimicking the low-density lipoprotein receptor. Thus, these NPs can transport cross the blood–brain barrier via the low-density lipoprotein receptor-mediated transcytosis due to the elevated expression of these receptors on the brain microvessel endothelium (Kreuter 2013). Moreover, there was an insignificant uptake of PMAA-PS80-*g*-starch NPs by the liver and by spleen, suggesting good targeting performance and less RES recognition of these NPs (Shalviri et al. 2012). The presence of PS80 on the particle surface allowed the adsorption of certain serum proteins such as proteins A-E, which limited the adsorption of hard corona proteins, hence changed the opsonization profile of the NPs and improved targeting efficiency of these NPs (Shalviri et al. 2012).

5.2.2.3 External Physical Targeting

Another possible strategy of targeting is the physical targeting of drugs with an external stimulus. Regarding this, magnetic field is a typical external stimulus for starch-based carriers (Chap. 4, Fig. 2). The magnetic capture of MNPs in tumor tissues occurs only when the magnetic force is stronger than the drag forces associated with the convective blood flow (Likhitkar and Bajpai 2014). To ensure a strong magnetic force generated on NPs, a large particle size is needed; however, it may also induce aggregation and low stability in the blood circulation (Cole et al. 2011b). To solve issues associated with MNPs, the surface coating of MNPs with starch could assist in the stabilization of MNPs, the control of the particle size, and the enhancement of biocompatibility and biodegradability (Bergemann et al. 1999; Darroudi et al. 2014; Kim et al. 2003; Lu et al. 2013).

Using starch derivatives as a coating material could provide an advantage of increasing the drug loading capacity. Starch can be grafted by functional polymer chains, which can hold drug molecules. For example, grafting starch-coated MNPs with negative heparin was shown capable of increasing the loading content of cationic drug proteins significantly (Zhang et al. 2013b, 2014). Delivery of hydrophobic drugs by magnetic carriers solved the problem of the insolubility of hydrophobic drugs in the blood medium and reduced the serious side effects caused by the cytotoxicity of anticancer drugs toward normal cells (Cole et al. 2011a; Jie et al. 2013). Furthermore, compared with MNPs, which often have relatively high biodistribution in both the liver (80–90% of dose) and the spleen (5–8% of dose) (Chertok et al. 2010; Lee et al. 2010), the coating of MNPs with biocompatible starch could reduce the RES uptake as well as the nonspecific interactions with plasma (Ali and AlArifi 2009; Fathi and Entezami 2014; Geresh et al. 2004; Rodrigues and Emeje 2012). However, MNPs coated with only starch still showed high RES recognition in the liver and the spleen and had a short plasma half-lifetime (Cole et al. 2011a, b; Jie et al. 2013; Zhang et al. 2013b, 2014). Thus, these starch-coated MNPs could have a low targeting efficiency.

Some efforts have been made to solve the problem (high RES recognition and low stability) of only starch-coated MNPs (Table 5.3). Further grafting starch with PEG (Cole et al. 2011a, b), PEG and heparin (Zhang et al. 2013b, 2014), and β-glucosidase (Jie et al. 2013) have been tried to improve the targeting efficiency of MNPs. These modified starch-coated MNPs showed a long circulating behavior in vivo and low RES recognition in the liver and the spleen or other macrophages. Specifically, the plasma half-lifetime of PEG and heparin-modified starch-coated MNPs was more than 60 folds longer than that of only starch-coated MNPs (Zhang et al. 2013b, 2014). For PEG (Cole et al. 2011a) and β-glucosidase (Jie et al. 2013) grafted starch-coated MNPs, 15-fold (1.0% injected dose Fe/g tissue) and 3.9-fold (289 ± 58 nmol Fe/g tissue) improvements have been achieved in targeting efficiency, respectively. Besides surface modification of starch-coated MNPs, starch magnetic NPs can also be encapsulated into various particulate carriers. A 27-fold increase of the drug half-lifetime in the blood plasma was achieved by incorporating starch-coated MNPs in liposomes (Nahar et al. 2014). An enhanced uptake of liposomes by the targeted cells (pulmonary arterial smooth muscle cells) was observed: 3-fold greater compared with that observed in the absence of EMF. Also, the pulmonary arterial smooth muscle cells proliferation was reduced by 40% under the influence of the magnetic field (Nahar et al. 2014).

5.2.3 DDSs That Target the Immune System

Due to the potential immunostimulatory, antigenic and adjuvant properties of drug-loaded particles (Behzadi et al. 2017), the uptake of particles by RESs is unavoidable. Thus, the tendency of starch-based micro- and nanocarriers to localize or activate the immune cells presents an opportunity for the targeting of bioactive molecules, such as vaccines and antigens, to the immune system. In such cases, therapeutics can be used to treat intracellular infections such as intracellular parasitic diseases (Laakso et al. 1986; Stjarnkvist et al. 1989, 1991) and leishmaniasis (Baillie et al. 1987), where situation phagocytic cells are directly involved in the disease process. Besides, another way to enhance the immune response to disease treatment is oral M cells-targeting delivery, and it has a huge advantage because of its ability of transportation of bioactive substances through M cells without degradation (Bye et al. 1984). Thus, the RES cells and M cells targeting of starch-based carriers will be also discussed regarding the ways to improve the effect and efficiency of these targeting DDSs.

Comprehensive understanding the effect of critical parameters (e.g., size and surface characteristics) of carriers on the uptake and activation of immune systems is necessary to further improve the treatment of immune-related diseases. The effect of parameters on the RES recognition and uptake follows the opposite trend as it behaves for the inhibition of opsonization (Table 5.3). Usually, the larger the size of starch-based carriers, the higher is the chance for the carriers to be recognized by RESs. Moreover, hydrophobic modification of starch-based carriers usually increases the chance of recognition by RESs (Table 5.3). For example, acryloylated starch MPs, prepared by radical polymerization of acryloylated starch ($DS \geq 2$), were observed to have a typical RES distribution (Laakso et al. 1986; Stjarnkvist et al. 1989, 1991). The half-lifetime of these MPs in the blood circulation was short (<5 min), and they were efficiently taken up by the RES system, mainly in the liver (50–90%) (Laakso et al. 1986). Still, there are mixed results, due to the complexity of biological interactions and the different experimental conditions. It was also reported that acryloylated starch MPs had low macrophage activating properties (Artursson et al. 1987). Also, the enhanced immune response can be achieved by the combination of starch-based carriers with targeting ligands/adjuvants at special immune cells (Table 5.3). HES microcapsules was equipped with monophosphoryl lipid A, anti-CD40, and anti-DEC205 to enhance liver deposition, to accelerate the uptake by non-parenchymal liver cells, and to induce the secretion of TNF-α, IL-6 by KCs, and IFN-γ and IL-12p70 by liver DCs (Fichter et al. 2015; Pietrzak-Nguyen et al. 2014).

Microorganisms and particles firstly adhere to the apical surface of M cells and are translocated through the cells. After the passage through the M cells, the particles/antigens are presented to the underlying cells of the immune system. Starch MPs have been successfully proven to help vaccine adjuvant [e.g., toxin B subunits (Larhed et al. 2004)], or proteins [e.g., HAS (Heritage et al. 1996, 1998; Larhed et al. 2004; McDermott et al. 1998)] delivered through M cells. For example, when conjugated with HSA, starch MPs were found in the follicles of the Peyer's patches and not in villi, suggesting the successful targeting of M cells (Larhed et al. 2004). Moreover, the combination grafted MPs with HSA dramatically enhanced serum IgG responses compared with soluble HSA alone and elicited HSA-specific IgA responses in gut secretions (Heritage et al. 1996, 1998; McDermott et al. 1998). Animal tests also showed that these HSA-loaded grafted MPs induced a shift of HSA-specific lymphocyte responsiveness from the mucosal toward the systemic compartment of the immune system (Heritage et al. 1996, 1998). Like the uptake of particles by RESs, among various factors, the particle size and hydrophobicity play the major role in the targeting of microspheres to M cells (Table 5.3). MPs in size range of 1–10 μm could be most effectively internalized by M cells. Also, hydrophobicization can help starch-based carriers to deliver antigens through M targeting delivery and induce an intense immune response. For example, acryloylated starch MPs combined with rotavirus vaccine induced strong and gradually increasing rotavirus-specific serum IgG responses after intramuscular immunization through M-targeting delivery (Strindelius et al. 2004; Sturesson and Degling Wikingsson 2000). Moreover, an increase in the hydrophobicity has a direct positive influence on the efficiency of phagocytosis (Chadha and DeLuca 1998; Sturesson and Degling Wikingsson 2000).

References

Aderem A, Underhill DM (1999) Mechanisms of phagocytosis in macrophages. Annu Rev Immunol 17:593–623

American Psychiatric A (1988) AIDS policy: guidelines for inpatient psychiatric units. Am J Psychiatry 145:542. https://doi.org/10.1155/2010/409320

Ahmad MZ, Akhter S, Anwar M, Ahmad FJ (2012a) Assam Bora rice starch based biocompatible mucoadhesive microsphere for targeted delivery of 5-fluorouracil in colorectal cancer. Mol Pharm 9:2986–2994. https://doi.org/10.1021/mp300289y

Ahmad MZ, Akhter S, Ahmad I, Singh A, Anwar M, Shamim M, Ahmad FJ (2012b) In vitro and in vivo evaluation of Assam Bora rice starch-based bioadhesive microsphere as a drug carrier for colon targeting. Expert Opin Drug Deliv 9:141–149. https://doi.org/10.1517/17425247.2012.633507

Ahmad MZ, Akhter S, Anwar M, Kumar A, Rahman M, Talasaz AH, Ahmad FJ (2013) Colorectal cancer targeted Irinotecan-Assam Bora rice starch based microspheres: a mechanistic, pharmacokinetic and biochemical investigation. Drug Dev Ind Pharm 39:1936–1943. https://doi.org/10.3109/03639045.2012.719906

Ali AE-H, AlArifi A (2009) Characterization and in vitro evaluation of starch based hydrogels as carriers for colon specific drug delivery systems. Carbohydr Polym 78:725–730

Anderberg EK, Nyström C, Artursson P (1992) Epithelial transport of drugs in cell culture. VII: Effects of pharmaceutical surfactant excipients and bile acids on transepithelial permeability in monolayers of human intestinal epithelial (Caco-2) cells. J Pharm Sci 81:879–887

Armstrong JK, Hempel G, Koling S, Chan LS, Fisher T, Meiselman HJ, Garratty G (2007) Antibody against poly (ethylene glycol) adversely affects PEG-asparaginase therapy in acute lymphoblastic leukemia patients. Cancer 110:103–111

Artursson P, Arro E, Edman P, Ericsson JL, Sjöholm I (1987) Biodegradable microspheres V: stimulation of macrophages with microparticles made of various polysaccharides. J Pharm Sci 76:127–133

Baier G, Baumann D, Siebert JM, Musyanovych A, Mailander V, Landfester K (2012) Suppressing unspecific cell uptake for targeted delivery using hydroxyethyl starch nanocapsules. Biomacromolecules 13:2704–2715. https://doi.org/10.1021/bm300653v

Baillie AJ, Coombs GH, Dolan TF, Hunter CA, Laakso T, Sjöholm I, Stjärnkvist P (1987) Biodegradable microspheres: polyacryl starch microparticles as a delivery system for the antileishmanial drug, sodium stibogluconate. J Pharm Pharmacol 39:832–835. https://doi.org/10.1111/j.2042-7158.1987.tb05126.x

Behzadi S et al (2017) Cellular uptake of nanoparticles: journey inside the cell. Chem Soc Rev 46:4218–4244. https://doi.org/10.1039/c6cs00636a

Bergemann C, Muller-Schulte D, Oster J, a Brassard L, Lubbe AS (1999) Magnetic ion-exchange nano- and microparticles for medical, biochemical and molecular biological applications. J Magn Magn Mater 194:45–52. https://doi.org/10.1016/s0304-8853(98)00554-x

Bie P, Chen L, Li X, Li L (2016) Characterization of concanavalin A-conjugated resistant starch acetate bioadhesive film for oral colon-targeting microcapsule delivery system. Ind Crops Prod 84:320–329. https://doi.org/10.1016/j.indcrop.2016.02.023

Biswas N, Sahoo RK (2016) Tapioca starch blended alginate mucoadhesive-floating beads for intragastric delivery of Metoprolol Tartrate. Int J Biol Macromol 83:61–70. https://doi.org/10.1016/j.ijbiomac.2015.11.039

Björk E, Edman P (1988) Degradable starch microspheres as a nasal delivery system for insulin. Int J Pharm 47:233–238. https://doi.org/10.1016/0378-5173(88)90236-0

Björk E, Edman P (1990) Characterization of degradable starch microspheres as a nasal delivery system for drugs. Int J Pharm 62:187–192. https://doi.org/10.1016/0378-5173(90)90232-S

Bjork E, Isaksson U, Edman P, Artursson P (1995) Starch microspheres induce pulsatile delivery of drugs and peptides across the epithelial barrier by reversible separation of the tight junctions. J Drug Target 2:501–507. https://doi.org/10.3109/10611869509015920

Boddupalli BM, Mohammed ZNK, Nath RA, Banji D (2010) Mucoadhesive drug delivery system: an overview. J Adv Pharm Technol Res 1:381–387. https://doi.org/10.4103/0110-5558.76436

Bye W, Allan C, Trier J (1984) Structure, distribution, and origin of M cells in Peyer's patches of mouse ileum. Gastroenterology 86:789–801

Campbell K, Craig DQ, McNally T (2008) Poly (ethylene glycol) layered silicate nanocomposites for retarded drug release prepared by hot-melt extrusion. Int J Pharm 363:126–131

Chadha H, DeLuca PP (1998) Effect of polytyrosine on the hydrophobicity of hydroxyethyl starch microspheres. Pharm Dev Technol 3:597–606

Chertok B, Cole AJ, David AE, Yang VC (2010) Comparison of electron spin resonance spectroscopy and inductively-coupled plasma optical emission spectroscopy for biodistribution analysis of iron-oxide nanoparticles. Mol Pharm 7:375–385

Codd JE, Deasy PB (1998) Formulation development and in vivo evaluation of a novel bioadhesive lozenge containing a synergistic combination of antifungal agents. Int J Pharm 173:13-24. https://doi.org/10.1016/s0378-5173(98)00228-2.

Cole AJ, David AE, Wang J, Galbán CJ, Yang VC (2011a) Magnetic brain tumor targeting and biodistribution of long-circulating PEG-modified, cross-linked starch-coated iron oxide nanoparticles. Biomaterials 32:6291–6301. https://doi.org/10.1016/j.biomaterials.2011.05.024

Cole AJ, David AE, Wang J, Galban CJ, Hill HL, Yang VC (2011b) Polyethylene glycol modified, cross-linked starch-coated iron oxide nanoparticles for enhanced magnetic tumor targeting. Biomaterials 32:2183–2193. https://doi.org/10.1016/j.biomaterials.2010.11.040
Constantin M, Fundueanu G, Cortesi R, Esposito E, Nastruzzi C (2003) Aminated polysaccharide microspheres as DNA delivery systems. Drug Delivery 10:139-149. https://doi.org/10.1080/10717540390215537
Critchley H, Davis SS, Farraj NF, Illum L (1994) Nasal absorption of desmopressin in rats and sheep. Effect of a bioadhesive microsphere delivery system. J Pharm Pharmacol 46:651–656
Dandekar P, Jain R, Stauner T, Loretz B, Koch M, Wenz G, Lehr C-M (2012) A hydrophobic starch polymer for nanoparticle-mediated delivery of docetaxel. Macromol Biosci 12:184–194. https://doi.org/10.1002/mabi.201100244
Darroudi M, Hakimi M, Goodarzi E, Oskuee RK (2014) Superparamagnetic iron oxide nanoparticles (SPIONs): green preparation, characterization and their cytotoxicity effects. Ceram Int 40:14641–14645. https://doi.org/10.1016/j.ceramint.2014.06.051
Dobrovolskaia MA, Aggarwal P, Hall JB, McNeil SE (2008) Preclinical studies to understand nanoparticle interaction with the immune system and its potential effects on nanoparticle biodistribution. Mol Pharm 5:487–495
Edman P, Björk E, Rydén L (1992) Microspheres as a nasal delivery system for peptide drugs. J Controlled Release 21:165–172. https://doi.org/10.1016/0168-3659(92)90018-M
Engelberth SA, Hempel N, Bergkvist M (2015) Chemically Modified Dendritic Starch: A Novel Nanomaterial for siRNA Delivery. Bioconjugate Chem 26:1766–1774
Fathi M, Entezami AA (2014) Stable aqueous dispersion of magnetic iron oxide core-shell nanoparticles prepared by biocompatible maleate polymers. Surf Interface Anal 46:145–151. https://doi.org/10.1002/sia.5362
Fichter M, Baier G, Dedters M, Pretsch L, Pietrzak-Nguyen A, Landfester K, Gehring S (2013) Nanocapsules generated out of a polymeric dexamethasone shell suppress the inflammatory response of liver macrophages. Nanomed Nanotechnol Biol Med 9:1223–1234. https://doi.org/10.1016/j.nano.2013.05.005
Fichter M et al (2015) Monophosphoryl lipid A coating of hydroxyethyl starch nanocapsules drastically increases uptake and maturation by dendritic cells while minimizing the adjuvant dosage. Vaccine 33:838–846. https://doi.org/10.1016/j.vaccine.2014.12.072
Frei E, Canellos GP (1980) Dose: a critical factor in cancer chemotherapy. Am J Med 69:585–594
Freichels H, Wagner M, Okwieka P, Meyer RG, Mailander V, Landfester K, Musyanovych A (2013) (Oligo)mannose functionalized hydroxyethyl starch nanocapsules: en route to drug delivery systems with targeting properties. J Mater Chem B 1:4338–4348. https://doi.org/10.1039/c3tb20138d
Geresh S, Gdalevsky GY, Gilboa I, Voorspoels J, Remon JP, Kost J (2004) Bioadhesive grafted starch copolymers as platforms for peroral drug delivery: a study of theophylline release. J Controlled Release 94:391–399
Goszczynski TM, Filip-Psurska B, Kempinska K, Wietrzyk J, Boratynski J (2014) Hydroxyethyl starch as an effective methotrexate carrier in anticancer therapy. Pharmacol Res & Perspect 2:e00047–e00047. https://doi.org/10.1002/prp2.47
Güler MA, Gök MK, Figen AK, Özgümüş S (2015) Swelling, mechanical and mucoadhesion properties of Mt/starch-g-PMAA nanocomposite hydrogels. Appl Clay Sci 112:44–52
Hallenbeck JM et al (1986) Polymorphonuclear leukocyte accumulation in brain regions with low blood flow during the early postischemic period. Stroke 17:246–253. https://doi.org/10.1161/01.str.17.2.246
Heritage PL, Loomes LM, Jianxiong J, Brook MA, Underdown BJ, McDermott MR (1996) Novel polymer-grafted starch microparticles for mucosal delivery of vaccines. Immunology 88:162–168. https://doi.org/10.1046/j.1365-2567.1996.d01-639.x
Heritage PL, Underdown BJ, Brook MA, McDermott MR (1998) Oral administration of polymer-grafted starch microparticles activates gut-associated lymphocytes and primes mice for a

subsequent systemic antigen challenge. Vaccine 16:2010–2017. https://doi.org/10.1016/s0264-410x(98)00085-1

Hoffmann S, Caysa H, Kuntsche J, Kreideweiß P, Leimert A, Mueller T, Mäder K (2013) Carbohydrate plasma expanders for passive tumor targeting: in vitro and in vivo studies. Carbohydr Polym 95:404–413

Höök F, Rodahl M, Kasemo B, Brzezinski P (1998) Structural changes in hemoglobin during adsorption to solid surfaces: effects of pH, ionic strength, and ligand binding. Proc Nat Acad Sci 95:12271–12276

Illum L, Jørgensen H, Bisgaard H, Krogsgaard O, Rossing N (1987) Bioadhesive microspheres as a potential nasal drug delivery system. Int J Pharm 39:189–199. https://doi.org/10.1016/0378-5173(87)90216-x

Illum L, Fisher AN, Jabbal-Gill I, Davis SS (2001) Bioadhesive starch microspheres and absorption enhancing agents act synergistically to enhance the nasal absorption of polypeptides. Int J Pharm 222:109–119. https://doi.org/10.1016/s0378-5173(01)00708-6

Jain AK, Khar RK, Ahmed FJ, Diwan PV (2008) Effective insulin delivery using starch nanoparticles as a potential trans-nasal mucoadhesive carrier. Eur J Pharm Biopharm 69:426-435. https://doi.org/10.1016/j.ejpb.2007.12.001

Jain S, Bajpai A (2013) Designing polyethylene glycol (PEG)–plasticized membranes of poly (vinyl alcohol-g-methyl methacrylate) and investigation of water sorption and blood compatibility behaviors. Des Monomers Polym 16:436–446

Jain R et al (2011) Enhanced cellular delivery of idarubicin by surface modification of propyl starch nanoparticles employing pteroic acid conjugated polyvinyl alcohol. Int J Pharm 420:147–155. https://doi.org/10.1016/j.ijpharm.2011.08.030

Jie Z, Jian Z, Allan ED, Victor CY (2013) Magnetic tumor targeting of β-glucosidase immobilized iron oxide nanoparticles. Nanotechnology 24:375102

Kelemen LE (2006) The role of folate receptor α in cancer development, progression and treatment: cause, consequence or innocent bystander? Int J Cancer 119:243–250

Kim DK, Zhang Y, Voit W, Rao KV, Muhammed M (2001) Synthesis and characterization of surfactant-coated superparamagnetic monodispersed iron oxide nanoparticles. J Magn Magn Mater 225:30–36. https://doi.org/10.1016/S0304-8853(00)01224-5

Kim DK, Mikhaylova M, Zhang Y, Muhammed M (2003) Protective coating of superparamagnetic iron oxide nanoparticles. Chem Mater 15:1617–1627. https://doi.org/10.1021/cm021349j

Kreuter J (2013) Mechanism of polymeric nanoparticle-based drug transport across the blood-brain barrier (BBB). J Microencapsul 30:49–54

Kumar P, Senthamilselvi S, Govindaraju M (2014) Phloroglucinol-encapsulated starch biopolymer: preparation, antioxidant and cytotoxic effects on HepG2 liver cancer cell lines. RSC Adv 4:26787–26795. https://doi.org/10.1039/c4ra02621g

Kumari S, Mg S, Mayor S (2010) Endocytosis unplugged: multiple ways to enter the cell. Cell Res 20:256. https://doi.org/10.1038/cr.2010.19

Kwag DS, Oh KT, Lee ES (2014) Facile synthesis of multilayered polysaccharidic vesicles. J Controlled Release 187:83–90. https://doi.org/10.1016/j.jconrel.2014.05.032

Laakso T, Artursson P, Sjoholm I (1986) Biodegradable microspheres IV: factors affecting the distribution and degradation of polyacryl starch microparticles. J Pharm Sci 75:962–967

Larhed A, Stertman L, Edvardsson E, Sjöholm I (2004) Starch microparticles as oral vaccine adjuvant: antigen-dependent uptake in mouse intestinal mucosa. J Drug Target 12:289–296. https://doi.org/10.1080/10611860420002236622

Lee MJ-E et al (2010) Rapid pharmacokinetic and biodistribution studies using cholorotoxin-conjugated iron oxide nanoparticles: a novel non-radioactive method. PLOS ONE 5:e9536. https://doi.org/10.1371/journal.pone.0009536

Lefnaoui S, Moulai-Mostefa N (2011) Formulation and in vitro evaluation of kappa-carrageenan-pregelatinized starch-based mucoadhesive gels containing miconazole. Starch - Stärke 63:512–521. https://doi.org/10.1002/star.201000141

Lefnaoui S, Moulai-Mostefa N (2014) Investigation and optimization of formulation factors of a hydrogel network based on kappa carrageenan–pregelatinized starch blend using an experimental design. Colloids Surf A 458:117–125. https://doi.org/10.1016/j.colsurfa.2014.01.007

Lemieux M, Bouchard F, Gosselin P, Paquin J, Mateescu MA (2011) The NCI-N87 cell line as a gastric epithelial barrier model for drug permeability assay. Biochem Biophys Res Commun 412:429–434

Lemieux M, Gosselin P, Mateescu MA (2015) Carboxymethyl starch mucoadhesive microspheres as gastroretentive dosage form. Int J Pharm 496:497–508

Li G et al (2014a) Hydroxyethyl starch conjugates for improving the stability, pharmacokinetic behavior and antitumor activity of 10-hydroxy camptothecin. Int J Pharm 471:234–244. https://doi.org/10.1016/j.ijpharm.2014.05.038

Li J et al (2014b) A multifunctional polymeric nanotheranostic system delivers doxorubicin and imaging agents across the blood-brain barrier targeting brain metastases of breast cancer. ACS Nano 8:9925–9940. https://doi.org/10.1021/nn501069c

Li K et al (2015) Mulberry-like dual-drug complicated nanocarriers assembled with apogossypolone amphiphilic starch micelles and doxorubicin hyaluronic acid nanoparticles for tumor combination and targeted therapy. Biomaterials 39:131–144. https://doi.org/10.1016/j.biomaterials.2014.10.073

Li G, Cai C, Qi Y, Tang X (2016) Hydroxyethyl starch–10-hydroxy camptothecin conjugate: synthesis, pharmacokinetics, cytotoxicity and pharmacodynamics research. Drug Deliv 23:277–284

Likhitkar S, Bajpai AK (2012) Magnetically controlled release of cisplatin from superparamagnetic starch nanoparticles. Carbohydr Polym 87:300–308. https://doi.org/10.1016/j.carbpol.2011.07.053

Likhitkar S, Bajpai AK (2014) An in vitro experimental approach to study magnetically targeted release of methotrexate from superparamagnetic starch nanocarriers. Int J Polym Mater Polym Biomater 63:941–950. https://doi.org/10.1080/00914037.2014.886232

Lin J, Zhang H, Chen Z, Zheng Y (2010) Penetration of lipid membranes by gold nanoparticles: insights into cellular uptake, cytotoxicity, and their relationship. ACS Nano 4:5421–5429

Liu Q, Yang X, Xu H, Pan K, Yang Y (2013) Novel nanomicelles originating from hydroxyethyl starch-g-polylactide and their release behavior of docetaxel modulated by the PLA chain length. Eur Polym J 49:3522-3529. https://doi.org/10.1016/j.eurpolymj.2013.08.012

Lu W, Shen Y, Xie A, Zhang W (2013) Preparation and protein immobilization of magnetic dialdehyde starch nanoparticles. J Phys Chem B 117:3720–3725. https://doi.org/10.1021/jp3110908

Madhav NS, Shakya AK, Shakya P, Singh K (2009) Orotransmucosal drug delivery systems: a review. J Controlled Release 140:2–11

Mailman D, Womack WA, Kvietys PR, Granger DN (1990) Villous motility and unstirred water layers in canine intestine. Am J Physiol 258:G238–G246. https://doi.org/10.1152/ajpgi.1990.258.2.G238

Mallick SP et al (2016) An in-depth analysis of the mechanical, electrical, and drug release properties of gelatin–starch phase-separated hydrogels. Polym Plast Technol Eng 55:1731–1742. https://doi.org/10.1080/03602559.2016.1171873

Mao SR, Chen JM, Wei ZP, Liu H, Bi DZ (2004) Intranasal administration of melatonin starch microspheres. Int J Pharm 272:37–43. https://doi.org/10.1016/j.ijpharm.2003.11.028

Marto J et al (2016) A quality by design (QbD) approach on starch-based nanocapsules: a promising platform for topical drug delivery. Colloids Surf B Biointerfaces 143:177–185. https://doi.org/10.1016/j.colsurfb.2016.03.039

McDermott MR, Heritage PL, Bartzoka V, Brook MA (1998) Polymer-grafted starch microparticles for oral and nasal immunization. Immunol Cell Biol 76:256–262. https://doi.org/10.1046/j.1440-1711.1998.00743.x

Minimol PF, Paul W, Sharma CP (2013) PEGylated starch acetate nanoparticles and its potential use for oral insulin delivery. Carbohydr Polym 95:1–8. https://doi.org/10.1016/j.carbpol.2013.02.021

Moghimi SM, Hunter AC, Murray JC (2001) Long-circulating and target-specific nanoparticles: theory to practice. Pharmacol Rev 53:283–318

Momeni A, Mohammadi MH (2009) Respiratory delivery of theophylline by size-targeted starch microspheres for treatment of asthma. J Microencapsul 26:701–710. https://doi.org/10.3109/02652040802685043

Montgomery PC, Rafferty DE (1998) Induction of secretory and serum antibody responses following oral administration of antigen with bioadhesive degradable starch microparticles. Oral Microbiol Immunol 13:139–149. https://doi.org/10.1111/j.1399-302X.1998.tb00725.x

Mortazavi SA, Smart JD (1993) An investigation into the role of water movement and mucus gel dehydration in mucoadhesion. J Controlled Release 25:197–203

Nahar K, Absar S, Patel B, Ahsan F (2014) Starch-coated magnetic liposomes as an inhalable carrier for accumulation of fasudil in the pulmonary vasculature. Int J Pharm 464:185–195. https://doi.org/10.1016/j.ijpharm.2014.01.007

Narayanan D, Nair S, Menon D (2015) A systematic evaluation of hydroxyethyl starch as a potential nanocarrier for parenteral drug delivery. Int J Biol Macromol 74:575–584. https://doi.org/10.1016/j.ijbiomac.2014.12.012

Newman JP (1987) Reaction to punishment in extraverts and psychopaths: Implications for the impulsive behavior of disinhibited individuals. J Res Pers 21:464–480

Noga M et al (2013) The effect of molar mass and degree of hydroxyethylation on the controlled shielding and deshielding of hydroxyethyl starch-coated polyplexes. Biomaterials 34:2530–2538. https://doi.org/10.1016/j.biomaterials.2012.12.025

Noga M, Edinger D, Wagner E, Winter G, Besheer A (2014) Characterization and compatibility of hydroxyethyl starch-polyethylenimine copolymers for DNA delivery. J Biomater Sci Polym Ed 25:855–871. https://doi.org/10.1080/09205063.2014.910152

O'Hagan DT, Rafferty D, Wharton S, Illum L (1993) Intravaginal immunization in sheep using a bioadhesive microsphere antigen delivery system Vaccine 11:660-664. https://doi.org/10.1016/0264-410X(93)90313-M

Oliveira Cardoso VM, Stringhetti Ferreira Cury B, Evangelista RC, Daflon Gremiao MP (2016) Development and characterization of cross-linked gellan gum and retrograded starch blend hydrogels for drug delivery applications. J Mech Behav Biomed Mater 65:317–333. https://doi.org/10.1016/j.jmbbm.2016.08.005

Paderni C, Compilato D, Giannola LI, Campisi G (2012) Oral local drug delivery and new perspectives in oral drug formulation. Oral Surg Oral Med Oral Pathol Oral Radiol 114:e25–e34. https://doi.org/10.1016/j.oooo.2012.02.016

Pietrzak-Nguyen A et al (2014) Enhanced in vivo targeting of murine nonparenchymal liver cells with monophosphoryl lipid A functionalized microcapsules. Biomacromolecules 15:2378–2388. https://doi.org/10.1021/bm5006728

Rajan M, Raj V (2013) Potential drug delivery applications of chitosan based nanomaterials. Int Rev Chem Eng 5:145–155

Reinhold AK, Rittner HL (2017) Barrier function in the peripheral and central nervous system—a review. Pflug Arch Eur J Phy 469:123–134. https://doi.org/10.1007/s00424-016-1920-8

Richardson JL, Farraj NF, Illum L (1992) Enhanced vaginal absorption of insulin in sheep using lysophosphatidylcholine and a bioadhesive microsphere delivery system. Int J Pharm 88:319–325. https://doi.org/10.1016/0378-5173(92)90330-5

Rodrigues A, Emeje M (2012) Recent applications of starch derivatives in nanodrug delivery. Carbohydr Polym 87:987–994. https://doi.org/10.1016/j.carbpol.2011.09.044

Roohi F, Lohrke J, Ide A, Schütz G, Dassler K (2012) Studying the effect of particle size and coating type on the blood kinetics of superparamagnetic iron oxide nanoparticles. Int J Nanomed 7:4447–4458. https://doi.org/10.2147/IJN.S33120

Ruge CA et al (2012) The interplay of lung surfactant proteins and lipids assimilates the macrophage clearance of nanoparticles. PLOS ONE 7:e40775. https://doi.org/10.1371/journal.pone.0040775

Saikia C, Hussain A, Ramteke A, Sharma HK, Maji TK (2014) Crosslinked thiolated starch coated Fe_3O_4 magnetic nanoparticles: effect of montmorillonite and crosslinking density on drug delivery properties. Starch - Stärke 66:760–771

Saikia C, Hussain A, Ramteke A, Sharma HK, Maji TK (2015) Carboxymethyl starch-chitosan-coated iron oxide magnetic nanoparticles for controlled delivery of isoniazid. J Microencapsul 32:29–39

Saikia C, Das MK, Ramteke A, Maji TK (2016) Effect of crosslinker on drug delivery properties of curcumin loaded starch coated iron oxide nanoparticles. Int J Biol Macromol 93:1121–1132. https://doi.org/10.1016/j.ijbiomac.2016.09.043

Saikia C, Das MK, Ramteke A, Maji TK (2017) Evaluation of folic acid tagged aminated starch/ZnO coated iron oxide nanoparticles as targeted curcumin delivery system. Carbohydr Polym 157:391–399. https://doi.org/10.1016/j.carbpol.2016.09.087

Satchi-Fainaro R, Duncan R, Barnes CM (2006) Polymer therapeutics for cancer: current status and future challenges. In: Satchi-Fainaro R, Duncan R (eds) Polymer therapeutics II, 1 edn. Springer, Berlin, pp 1–65. https://doi.org/10.1007/12_024

Shalviri A, Cai P, Rauth AM, Henderson JT, Wu XY (2012) Evaluation of new bi-functional terpolymeric nanoparticles for simultaneous in vivo optical imaging and chemotherapy of breast cancer. Drug Delivery Transl Res 2:437–453. https://doi.org/10.1007/s13346-012-0103-1

Sieradzki R, Traitel T, Goldbart R, Geresh S, Kost J (2014) Tailoring quaternized starch as a non-viral carrier for gene delivery applications. Polym Adv Technol 25:552–561. https://doi.org/10.1002/pat.3277

Situ W, Li X, Liu J, Chen L (2015) Preparation and characterization of glycoprotein-resistant starch complex as a coating material for oral bioadhesive microparticles for colon-targeted polypeptide delivery. J Agric Food Chem 63:4138–4147

Soane RJ, Frier M, Perkins AC, Jones NS, Davis SS, Illum L (1999) Evaluation of the clearance characteristics of bioadhesive systems in humans. Int J Pharm 178:55–65. https://doi.org/10.1016/s0378-5173(98)00367-6

Stjarnkvist P, Laakso T, Sjoholm I (1989) Biodegradable microspheres XII: properties of the crosslinking chains in polyacryl starch microparticles. J Pharm Sci 78:52–56

Stjarnkvist P, Degling L, Sjoholm I (1991) Biodegradable microspheres. XIII: immune response to the DNP hapten conjugated to polyacryl starch microparticles. J Pharm Sci 80:436–440

Strindelius L, Folkesson A, Normark S, Sjoholm I (2004) Immunogenic properties of the Salmonella atypical fimbriae in BALB/c mice. Vaccine 22:1448–1456. https://doi.org/10.1016/j.vaccine.2003.10.012

Sturesson C, Degling Wikingsson L (2000) Comparison of poly(acryl starch) and poly(lactide-co-glycolide) microspheres as drug delivery system for a rotavirus vaccine. J Controlled Release 68:441–450. https://doi.org/10.1016/s0168-3659(00)00294-7

Surini S, Anggriani V, Anwar E (2009) Study of mucoadhesive microspheres based on pregelatinized cassava starch succinate as a new carrier for drug delivery. J Med Sci 9:249–256. https://doi.org/10.3923/jms.2009.249.256

Thakore S, Valodkar M, Soni JY, Vyas K, Jadeja RN, Devkar RV, Rathore PS (2013) Synthesis and cytotoxicity evaluation of novel acylated starch nanoparticles. Bioorganic Chem 46:26–30. https://doi.org/10.1016/j.bioorg.2012.10.001

Thiele C, Auerbach D, Jung G, Qiong L, Schneider M, Wenz G (2011) Nanoparticles of anionic starch and cationic cyclodextrin derivatives for the targeted delivery of drugs. Polym Chem 2:209–215. https://doi.org/10.1039/c0py00241k

Vasir JK, Tambwekar K, Garg S (2003) Bioadhesive microspheres as a controlled drug delivery system. Int J Pharm 255:13–32. https://doi.org/10.1016/S0378-5173(03)00087-5

Vyas SP, Jain CP (1992) Bioadhesive polymer-grafted starch microspheres bearing isosorbide dinitrate for buccal administration. J Microencapsul 9:457–464. https://doi.org/10.3109/02652049209040484

Wang J, Liu H, Leng F, Zheng L, Yang J, Wang W, Huang CZ (2014) Autofluorescent and pH-responsive mesoporous silica for cancer-targeted and controlled drug release. Microporous Mesoporous Mater 186:187–193. https://doi.org/10.1016/j.micromeso.2013.11.006
Winzen S, Schoettler S, Baier G, Rosenauer C, Mailaender V, Landfester K, Mohr K (2015) Complementary analysis of the hard and soft protein corona: sample preparation critically effects corona composition. Nanoscale 7:2992–3001. https://doi.org/10.1039/c4nr05982d
Witschi C, Mrsny RJ (1999) In vitro evaluation of microparticles and polymer gels for use as nasal platforms for protein delivery. Pharm Res 16:382–390. https://doi.org/10.1023/a:1018869601502
Xiao SY et al (2006) Preparation of folate-conjugated starch nanoparticles and its application to tumor-targeted drug delivery vector. Chin Sci Bull 51:1693–1697. https://doi.org/10.1007/s11434-006-2039-7
Xiao H, Yang T, Lin Q, Liu GQ, Zhang L, Yu F, Chen Y (2016) Acetylated starch nanocrystals: preparation and antitumor drug delivery study. Int J Biol Macromol 89:456–464. https://doi.org/10.1016/j.ijbiomac.2016.04.037
Yadav AV, Mote HH (2008) Development of Biodegradable Starch Microspheres for Intranasal Delivery. Indian J Pharm Sci 70:170–174. https://doi.org/10.4103/0250-474X.41450
Yamada H, Loretz B, Lehr C-M (2014) Design of starch-graft-PEI polymers: an effective and biodegradable gene delivery platform. Biomacromolecules 15:1753–1761
Yang Y, Jiang JS, Du B, Gan ZF, Qian M, Zhang P (2009) Preparation and properties of a novel drug delivery system with both magnetic and biomolecular targeting. J Mater Sci Mater Med 20:301–307. https://doi.org/10.1007/s10856-008-3577-0
Ye L et al (2016) Zwitterionic-modified starch-based stealth micelles for prolonging circulation time and reducing macrophage response. ACS Appl Mater Interfaces 8:4385–4398
Zhang AP et al (2013a) Disulfide crosslinked PEGylated starch micelles as efficient intracellular drug delivery platforms. Soft Matter 9:2224–2233. https://doi.org/10.1039/c2sm27189c
Zhang J, Shin MC, David AE, Zhou J, Lee K, He H, Yang VC (2013b) Long-circulating heparin-functionalized magnetic nanoparticles for potential application as a protein drug delivery platform. Mol Pharm 10:3892–3902. https://doi.org/10.1021/mp400360q
Zhang L et al (2013c) Zwitterionic hydrogels implanted in mice resist the foreign-body reaction. Nat Biotechnol 31:553–556
Zhang J, Shin MC, Yang VC (2014) Magnetic targeting of novel heparinized iron oxide nanoparticles evaluated in a 9L-glioma mouse model. Pharm Res 31:579–592. https://doi.org/10.1007/s11095-013-1182-5

Chapter 6
Toxicology of Starch-Based DDSs

Abstract Because of its natural source and high biocompatibility, starch has drawn much attention in the advance of drug delivery systems (DDSs) as carrier materials. Yet, to functionalize starch to meet the material needs of smart and well-performed DDSs, starch has to be modified by numerous chemical methods, which may hamper its biocompatibility and biodegradability. Thus, it is vital to evaluate the toxicology of starch-based DDSs for its clinical application. Generally, starch-based DDSs could guarantee high biocompatibility and biodegradation. In most cases, the cellular toxicity of starch-based nanoparticulate carriers showed none or low toxicity and good compatibility with normal cells. However, in particular situations, a significant level of cytotoxicity of starch-based DDSs still exist. The toxicity of starch-based DDSs was found to be dosage-dependent, be related to the physicochemical properties such as size, charge, shape, molecular mass (M), composition, and chemical properties. Moreover, the addition of inorganic materials may bring cytotoxicity to starch-based DDSs, and its content is vital for its potential cytotoxicity. Thus, much effort is needed to create safer starch-based DDSs, despite the fact that starch still has a huge advantage over synthesized polymers in the application of drug delivery.

Keywords Starch-based carrier material · Biocompatibility · Biodegradability · Dosage-dependent toxicity · Chemical composition · Surface hydrophobicity · Surface charge

The renewable, biocompatible, and biodegradable starch has been used vastly to lower the toxicity of drug delivery carriers. The evaluation of the *in vitro* cytotoxicity of a biomaterial is the initial step of biocompatibility study. Normally, starch-based DDSs could guarantee high biocompatibility and biodegradation. In most cases, the cellular toxicity of starch-based nanoparticulate carriers showed none or low toxicity and good compatibility with normal cells (Athira and Jyothi 2015; Chen et al. 2016a, b; Devy et al. 2006; Engelberth et al. 2015; Huang et al. 2013; Hubatsch et al. 2007; Jain et al. 2011; Kwag et al. 2014; Li et al. 2015; Liu et al. 2015; Noga et al. 2013, 2014; Pereira et al. 2013; Shalviri et al. 2012; Tuovinen et al. 2004; Wang et al. 2014; Xiao et al. 2016; Yang et al. 2014b; Ye et al. 2016; Ying et al. 2014; Zhang et al. 2013). This is also true for starch MNPs (Fathi et al. 2015; Khan et al. 2012; Likhitkar and Bajpai 2014; Lima-Tenório et al. 2015; Nahar et al. 2014; Pourjavadi et al. 2015;

J. Chen et al., *Drug Delivery Applications of Starch Biopolymer Derivatives*
https://doi.org/10.1007/978-981-13-3657-7_6

Saikia et al. 2015, 2016; Xuan Phuc et al. 2012; Yang et al. 2009; Zohreh et al. 2016). Nontoxic effect of starch-coated MNPs was observed even in concentrations up to 62.5 μg/mL (Darroudi et al. 2014). However, in particular situations, a significant level of cytotoxicity of starch-based DDSs may still exist. The cytotoxicity of starch-based DDSs was shown to be dosage-dependent (Thakore et al. 2013). For example, the drug-loaded starch NPs with disulfide bonds as the crosslinks was noncytotoxic at low concentrations (0.003–0.3 μg/mL). However, a slight cytotoxicity was observed at the concentrations (≥30 μg/mL) of these starch NPs (Yang et al. 2014a). Also, the cytotoxicity of starch-based DDSs depends on the duration time of application. For example, while DSMs did not induce erythrocyte hemolysis initially, 8 weeks after administration of DSMs, the nasal septum showed mild, focal hyperplasia of the epithelium (Björk and Edman 1990).

When using nanoparticle for injection, its toxicity relies on multiple factors and can be regulated by a variety of physicochemical properties such as the size, charge, shape, *M*, composition, and chemical properties of starch-based polymers (Krug and Wick 2011). For example, the cytotoxic activity of acylated starch NPs was found to be dosage-dependent at a high dosage range of 1000–10,000 μg/mL, which may be due to the different hydrophobicity, and types of substitution (Thakore et al. 2013). Specifically, the DS of acylated starch NPs was ≥2, indicating the hydrophobic nature of acylated starch. This hydrophobicity may be responsible for the cytotoxicity of NPs (Thakore et al. 2013). Moreover, concerning the type of substitution to acylated starch NPs, starch-cinnamate NPs showed the highest cytotoxicity and nearly 80% cytotoxicity was recorded with its highest dosage (10,000 μg/mL). Nonetheless, starch-palmitate NPs was recorded with a low cytotoxic potential, probably because of its completely aliphatic structure (Thakore et al. 2013). Furthermore, for PEG-modified starches ($SA\text{-}PEG_{1900}$, $SA\text{-}PEG_{800}$, and $SA\text{-}PEG_{500}$ micelles, with their *M*'s being 1900, 800, and 500, respectively), the PEG of a low *M* was shown to be more toxic than the PEG of a high *M*. The starch-PEG samples with a low *M* ($SA\text{-}PEG_{800}$ and $SA\text{-}PEG_{500}$) were toxic with a cell viability less than 10% of viable cells. Moreover, the hemolysis caused by $SA\text{-}PEG_{1900}$ was practically nil compared to the other two derivatives, which were in the order of 4% (Minimol et al. 2013). Thus, when using PEG to modulate the properties of starch, the *M* of PEG need to be considered for the application of PEGlated starch.

The addition of inorganic materials may bring cytotoxicity to starch-based DDSs. The cell cytotoxicity test of CMS-chitosan-coated MNPs/MMT showed that the cell viability was increased with the increased MMT concentration up to 3%. Beyond this concentration, the cell viability was decreased (Saikia et al. 2015). It also found that the cell viability of normal model cells was reduced up to 61% with 0.5% ZnO concentration of aminated starch/ZnO-coated MNPs, and was further reduced up to 58% by incorporating FA into the system (Saikia et al. 2017). These results indicate that when inorganic materials are incorporated to starch-based DDSs, the contents of them are critical to the cytotoxicity of the systems.

While the certain charge of cationic starch polymers is necessary for the high transfection for gene delivery systems, it may also induce high cytotoxicity to cells. For example, starch MPs prepared by chemically cross-linking, and then aminated with

N,N-diethyl-2-chloroethylamine or *N*-glycidyl-*N,N*-dimethyl-*N*-methylammonium chloride, showed a moderate cytotoxic effect [half maximal inhibitory concentration (IC_{50}) values: 62.5 μM] (Constantin et al. 2003).

To endow starch with versatile functions to satisfy the material needs of smart and well-performed DDSs, starch has to be modified by numerous methods, which may jeopardize its biocompatibility and biodegradability. Nevertheless, in most cases, the nontoxicity or low toxicity of starch remains. Thus, starch still has a huge advantage over synthesized polymers in the application of drug delivery. Despite that, efforts have been made to create safer starch-based DDSs. For example, to address the toxicity issue of cationic starch polymers, a biodegradable starch-based polymer with tunable charge density by grafting short diamine side chains onto starch has been prepared to achieve efficient delivery of DNA with a much safer administration. This type of starch NP was shown to be noncytotoxic even at rather high concentrations ($IC_{50} > 400$ μg/ml) (Thiele et al. 2017).

References

Athira GK, Jyothi AN (2015) Cassava starch-poly (vinyl alcohol) nanocomposites for the controlled delivery of curcumin in cancer prevention and treatment. Starch-Stärke 67:549–558

Björk E, Edman P (1990) Characterization of degradable starch microspheres as a nasal delivery system for drugs. Int J Pharm 62:187–192. https://doi.org/10.1016/0378-5173(90)90232-S

Chen M, Gao C, Lü S, Chen Y, Liu M (2016a) Dual redox-triggered shell-sheddable micelles self-assembled from mPEGylated starch conjugates for rapid drug release. RSC Adv 6:9164–9174

Chen M, Gao C, Lü S, Chen Y, Liu M (2016b) Preparation of redox-sensitive, core-crosslinked micelles self-assembled from mPEGylated starch conjugates: remarkable extracellular stability and rapid intracellular drug release. RSC Adv 6:46159–46169

Constantin M, Fundueanu G, Cortesi R, Esposito E, Nastruzzi C (2003) Aminated polysaccharide microspheres as DNA delivery systems. Drug Delivery 10:139–149. https://doi.org/10.1080/10717540390215537

Darroudi M, Hakimi M, Goodarzi E, Oskuee RK (2014) Superparamagnetic iron oxide nanoparticles (SPIONs): green preparation, characterization and their cytotoxicity effects. Ceram Int 40:14641–14645 https://doi.org/10.1016/j.ceramint.2014.06.051

Devy J, Balasse E, Kaplan H, Madoulet C, Andry MC (2006) Hydroxyethylstarch microcapsules: A preliminary study for tumor immunotherapy application. Int J Pharm 307:194–200. https://doi.org/10.1016/j.ijpharm.2005.09.035

Engelberth SA, Hempel N, Bergkvist M (2015) Chemically modified dendritic starch: a novel nanomaterial for siRNA delivery. Bioconjug Chem 26:1766–1774

Fathi M, Entezami AA, Arami S, Rashidi M-R (2015) Preparation of N-isopropylacrylamide/itaconic acid magnetic nanohydrogels by modified starch as a crosslinker for anticancer drug carriers. Int J Polym Mater Polym Biomater 64:541–549 https://doi.org/10.1080/00914037.2014.996703

Huang Y, Liu M, Gao C, Yang J, Zhang X, Zhang X, Liu Z (2013) Ultra-small and innocuous cationic starch nanospheres: preparation, characterization and drug delivery study. Int J Biol Macromol 58:231–239 doi:https://doi.org/10.1016/j.ijbiomac.2013.04.006

Hubatsch I, Ragnarsson EG, Artursson P (2007) Determination of drug permeability and prediction of drug absorption in Caco-2 monolayers. Nat Protoc 2:2111–2119

Jain R et al (2011) Enhanced cellular delivery of idarubicin by surface modification of propyl starch nanoparticles employing pteroic acid conjugated polyvinyl alcohol. Int J Pharm 420:147–155. https://doi.org/10.1016/j.ijpharm.2011.08.030

Khan A, El-Toni AM, Alhoshan M (2012) Preparation of thermo-responsive hydrogel-coated magnetic nanoparticles. Mater Lett 89:12–15. https://doi.org/10.1016/j.matlet.2012.08.064

Krug HF, Wick P (2011) Nanotoxicology: an interdisciplinary challenge. Angew Chem Int Ed 50:1260–1278

Kwag DS, Oh KT, Lee ES (2014) Facile synthesis of multilayered polysaccharidic vesicles. J Control Release 187:83–90. https://doi.org/10.1016/j.jconrel.2014.05.032

Li K et al (2015) Mulberry-like dual-drug complicated nanocarriers assembled with apogossypolone amphiphilic starch micelles and doxorubicin hyaluronic acid nanoparticles for tumor combination and targeted therapy. Biomaterials 39:131–144. https://doi.org/10.1016/j.biomaterials.2014.10.073

Likhitkar S, Bajpai AK (2014) An in vitro experimental approach to study magnetically targeted release of methotrexate from superparamagnetic starch nanocarriers. Int J Polym Mater Polymeric Biomater 63:941–950. https://doi.org/10.1080/00914037.2014.886232

Lima-Tenório MK, Tenório-Neto ET, Guilherme MR, Garcia FP, Nakamura CV, Pineda EAG, Rubira AF (2015) Water transport properties through starch-based hydrogel nanocomposites responding to both pH and a remote magnetic field. Chem Eng J 259:620–629. https://doi.org/10.1016/j.cej.2014.08.045

Liu K, Wang Y, Li H, Duan Y (2015) A facile one-pot synthesis of starch functionalized graphene as nano-carrier for pH sensitive and starch-mediated drug delivery. Colloids Surf B 128:86–93. https://doi.org/10.1016/j.colsurfb.2015.02.010

Minimol PF, Paul W, Sharma CP (2013) PEGylated starch acetate nanoparticles and its potential use for oral insulin delivery. Carbohyd Polym 95:1–8. https://doi.org/10.1016/j.carbpol.2013.02.021

Nahar K, Absar S, Patel B, Ahsan F (2014) Starch-coated magnetic liposomes as an inhalable carrier for accumulation of fasudil in the pulmonary vasculature. Int J Pharm 464:185–195. https://doi.org/10.1016/j.ijpharm.2014.01.007

Noga M et al (2013) The effect of molar mass and degree of hydroxyethylation on the controlled shielding and deshielding of hydroxyethyl starch-coated polyplexes. Biomaterials 34:2530–2538. https://doi.org/10.1016/j.biomaterials.2012.12.025

Noga M, Edinger D, Wagner E, Winter G, Besheer A (2014) Characterization and compatibility of hydroxyethyl starch-polyethylenimine copolymers for DNA delivery. J Biomater Sci Polym Ed 25:855–871. https://doi.org/10.1080/09205063.2014.910152

Pereira AGB, Fajardo AR, Nocchi S, Nakamura CV, Rubira AF, Muniz EC (2013) Starch-based microspheres for sustained-release of curcumin: preparation and cytotoxic effect on tumor cells. Carbohyd Polym 98:711–720. https://doi.org/10.1016/j.carbpol.2013.06.013

Pourjavadi A, Tehrani ZM, Hosseini SH (2015) Dendritic magnetite decorated by pH-responsive PEGylated starch: a smart multifunctional nanocarrier for the triggered release of anti-cancer drugs. RSC Adv 5:48586–48595

Saikia C, Hussain A, Ramteke A, Sharma HK, Maji TK (2015) Carboxymethyl starch-chitosan-coated iron oxide magnetic nanoparticles for controlled delivery of isoniazid. J Microencapsul 32:29–39

Saikia C, Das MK, Ramteke A, Maji TK (2016) Effect of crosslinker on drug delivery properties of curcumin loaded starch coated iron oxide nanoparticles. Int J Biol Macromol 93:1121–1132. https://doi.org/10.1016/j.ijbiomac.2016.09.043

Saikia C, Das MK, Ramteke A, Maji TK (2017) Evaluation of folic acid tagged aminated starch/ZnO coated iron oxide nanoparticles as targeted curcumin delivery system. Carbohyd Polym 157:391–399. https://doi.org/10.1016/j.carbpol.2016.09.087

Shalviri A et al. (2012) pH-Dependent doxorubicin release from terpolymer of starch, polymethacrylic acid and polysorbate 80 nanoparticles for overcoming multi-drug resistance in human

breast cancer cells. Eur J Pharm Biopharm 82:587–597 https://doi.org/10.1016/j.ejpb.2012.09.001
Thakore S, Valodkar M, Soni JY, Vyas K, Jadeja RN, Devkar RV, Rathore PS (2013) Synthesis and cytotoxicity evaluation of novel acylated starch nanoparticles. Bioorg Chem 46:26–30. https://doi.org/10.1016/j.bioorg.2012.10.001
Thiele C, Loretz B, Lehr C-M (2017) Biodegradable starch derivatives with tunable charge density—synthesis, characterization, and transfection efficiency. Drug Deliv Transl Res 7:252–258 https://doi.org/10.1007/s13346-016-0333-8
Tuovinen L et al (2004) Starch acetate microparticles for drug delivery into retinal pigment epithelium - in vitro study. J Controlled Release 98:407–413. https://doi.org/10.1016/j.jconrel.2004.05.016
Wang J, Liu H, Leng F, Zheng L, Yang J, Wang W, Huang CZ (2014) Autofluorescent and pH-responsive mesoporous silica for cancer-targeted and controlled drug release. Microporous Mesoporous Mater 186:187–193. https://doi.org/10.1016/j.micromeso.2013.11.006
Xiao H, Yang T, Lin Q, Liu GQ, Zhang L, Yu F, Chen Y (2016) Acetylated starch nanocrystals: Preparation and antitumor drug delivery study. Int J Biol Macromol 89:456–464. https://doi.org/10.1016/j.ijbiomac.2016.04.037
Xuan Phuc N et al (2012) Iron oxide-based conjugates for cancer theragnostics. Adv Nat Sci Nanosci Nanotechnol 3:033001. https://doi.org/10.1088/2043-6262/3/3/033001
Yang Y, Jiang JS, Du B, Gan ZF, Qian M, Zhang P (2009) Preparation and properties of a novel drug delivery system with both magnetic and biomolecular targeting. J Mater Sci - Mater Med 20:301–307. https://doi.org/10.1007/s10856-008-3577-0
Yang J, Huang Y, Gao C, Liu M, Zhang X (2014a) Fabrication and evaluation of the novel reduction-sensitive starch nanoparticles for controlled drug release. Colloids Surf B 115:368–376. https://doi.org/10.1016/j.colsurfb.2013.12.007
Yang JL, Gao CM, Lu SY, Wang XG, Chen MJ, Liu MZ (2014b) Novel self-assembled amphiphilic mPEGylated starch-deoxycholic acid polymeric micelles with pH-response for anticancer drug delivery. RSC Adv 4:55139–55149. https://doi.org/10.1039/c4ra07315k
Ye L et al (2016) Zwitterionic-modified starch-based stealth micelles for prolonging circulation time and reducing macrophage response. ACS Appl Mater Interfaces 8:4385–4398
Ying XY, Shan CL, Jiang KK, Chen Z, Du YZ (2014) Intracellular pH-sensitive delivery $CaCO_3$ nanoparticles templated by hydrophobic modified starch micelles. RSC Adv 4:10841–10844 https://doi.org/10.1039/c3ra47501h
Zhang AP et al (2013) Disulfide crosslinked PEGylated starch micelles as efficient intracellular drug delivery platforms. Soft Matter 9:2224–2233. https://doi.org/10.1039/c2sm27189c
Zohreh N, Hosseini SH, Pourjavadi A (2016) Hydrazine-modified starch coated magnetic nanoparticles as an effective pH-responsive nanocarrier for doxorubicin delivery. J Ind Eng Chem 39:203–209. https://doi.org/10.1016/j.jiec.2016.05.029

Chapter 7
Conclusion and Future Perspectives

Abstract Starch from various natural origins and its derivatives are well known to be biodegradable and biocompatible and thus could be promising materials to construct next-generation drug delivery systems (DDSs). With a better understanding of the human physiological basis, advanced stimulus-responsive, target-specific, and transmucosal DDSs has been proven to be effective to deliver therapeutics to various sites in the body to maximize therapeutic potential and minimize side effects. Starch has been modified by various chemical, physical, and/or enzymatic methods to meet the specific requirements of these different DDSs. Yet, so far, starch is not as versatile as synthetic polymers for the development of DDSs, and the field on starch-based DDSs still needs the improvement of the currently developed systems. Future improvements of starch-based DDSs can be achieved on three aspects: (a) to develop modified methods of starch to more greenness, such as physical and enzymatic methods, and genetic engineering technology; (b) to advance starch-based DDSs with new stimulus-responsiveness and high targeting efficiency, such as active targeting delivery; (c) to deepen the research on the plasma hematocompatibility and in vivo distribution patterns of starch-based carriers, and to establish the good in vitro–in vivo correlation of starch-based DDSs for better clinical application. More effort is needed to enlarge the use of starch-based DDSs and its clinical application potential.

Keywords Stimulus-responsive starch-based drug delivery · Target-specific starch-based drug delivery · Green modified methods · Hematocompatibility · in vivo distribution pattern · In vitro–in vivo correlation

The design, construction, and testing of smart starch-based vehicles have undergone a veritable explosion of interest in recent years. Starch from various natural origins and its derivatives are claimed to be the promising materials to construct next-generation DDSs. There are several possible motivations for the efforts of the starch application in drug delivery. First, good biocompatibility and nontoxicity will be brought to DDSs by using starch as the carrier material, along with additional advantages such as the low cost, and high availability of starch. Furthermore, starch, due to its special chemical structure, can be easily modified, resulting in varied enzymatic resistance and diverse starch-based functional materials to meet specific requirements of differ-

J. Chen et al., *Drug Delivery Applications of Starch Biopolymer Derivatives*
https://doi.org/10.1007/978-981-13-3657-7_7

ent DDSs. Yet, up to date, most modifications of starch have been performed chemically. Only limited studies have dealt with the combination of physical and enzymatic methods, which are much safer to humans than chemically modified starches. On this behalf, more efforts on new, especially "green," starch modification methods need to be dedicated to reduce or eliminate the use of environmentally and human hazardous chemicals. As an example, genetic engineering technologies have a high potential to create new starch cultivars by targeting enzymes for starch biosynthesis pathways in plants (Zeeman et al. 2010). In this way, the raw starch may have an excellent application prospect to address the challenges of multifunctional DDSs.

A better understanding of physiological differences at the organ and cell levels could allow us to develop starch-based DDSs with stimulus responsiveness, thanks to the advanced material design. Starch-based DDSs have been successfully developed to be responsive to stimuli such as pH, enzyme, temperature, and magnetic field for the delivery of drugs, genes, proteins, and vaccines. Moreover, by taking advantage of biological interactions between materials and tissues, cells, or biomacromolecules, target-specific, and bioadhesive properties have been endowed with starch-based DDSs to increase drug residence at the targeted sites and to improve enhance bioavailability of drugs. By modulating the parameters through the EPR effect or coupling active targeting ligands to starch-based carriers especially nanocarriers, the targeting efficacy of starch-based DDSs can be achieved by ways of passive, active, or physical targeting.

The application of starch in smart DDSs has made some success, yet, there are still lots of space for improvement. First, starch is not as versatile as synthetic polymers for the development of DDSs. As a stimulus-responsive polymer, up to date, starch and its derivatives have been developed to respond to the narrow range of stimuli. Other stimuli, such as hypoxia, nucleic acids, and mechanical cues in blood vessels that are associated with various diseases including cancer, cardiomyopathy, and vascular diseases (Lu et al. 2014), have not been exploited in starch-based DDSs. Moreover, the literature contains limited data on the key studies that affect the high performance of DDSs and the successful clinical application. Despite the advances in targeting delivery, the design of starch-based materials that can actively target-specific cells in the body remains to be challenged. Rare studies have focused on active targeting delivery, plasma hematocompatibility, and in vivo distribution pattern at a cellular level or in the RES tissues, which affect both their efficacy and safety of target-specific NPs platforms. Furthermore, for the benefit of patients, more understanding of how disease physiology affects biological activities is highly desired for the good establishment of in vitro–in vivo correlation and the better clinical application of starch-based DDSs. The significant difference in the pathophysiological conditions between patients and healthy subjects, such as the abnormal pH of the colon (Fallingborg et al. 1993; Sasaki et al. 1997), can lead to the failure of pH-sensitive colon-specific starch-based DDS in patients. More efforts are needed to enlarge the use of starch-based DDSs with the development of starch polymers with the expanded functionality and with a better understanding of the biological process in human. There will be no slowdown in starch and chemical engineering scientific efforts to continuously improve the performance of starch, and new advances in smart drug delivery system are foreseen.

Acknowledgements This article has been financially supported by the National Key R&D Program of China (no. 2016YFD0400203), the NSFC (31871751), The key project of Guangzhou Science and Technology Program (No.201804020036) and YangFan Innovative and Entrepreneurial Research Team Project (2014YT02S029).

References

Fallingborg J, Christensen LA, Jacobsen BA, Rasmussen SN (1993) Very low intraluminal colonic pH in patients with active ulcerative colitis. Dig Dis Sci 38:1989–1993

Lu Y, Sun W, Gu Z (2014) Stimuli-responsive nanomaterials for therapeutic protein delivery. J Control Release 194:1–19. https://doi.org/10.1016/j.jconrel.2014.08.015

Sasaki Y, Hada R, Nakajima H, Fukuda S, Munakata A (1997) Improved localizing method of radiopill in measurement of entire gastrointestinal pH profiles: colonic luminal pH in normal subjects and patients with Crohn's disease. Am J Gastroenterol 92

Zeeman SC, Kossmann J, Smith AM (2010) Starch: its metabolism, evolution, and biotechnological modification in plants. Annu Rev Plant Biol 61:209–234. https://doi.org/10.1146/annurev-arplant-042809-112301